T0116219

Praise for

Gregory Chaitin

and

PROVING DARWIN

"[Chaitin is] enthusiastic and extremely readable. . . . Amazing. No wonder [he] chose mathematics over physics." —*New Scientist*

"The musings of a great mind on a subject beyond most of us." —*Ottawa Citizen*

"Provocative. . . . Stimulating. . . . Credit the author for a lively style, lots of useful historical references and an appendix that includes von Neumann's prescient essay on self-reproducing automata." —*Kirkus Reviews*

"Gregory Chaitin has devoted his life to the study of mathematics. . . . A great mind."
  —*The Victoria Times Colonist*

"[Chaitin] is a creative genius." —*Tucson Citizen*

Also by Gregory Chaitin

Meta Math! The Quest for Omega
Algorithmic Information Theory
Conversations with a Mathematician
Exploring Randomness
From Philosophy to Program Size
Information, Randomness and Incompleteness
Information-Theoretic Incompleteness
Mathematics, Complexity and Philosophy
The Limits of Mathematics
The Unknowable
Thinking about Gödel and Turing
With Ugo Pagallo: *Teoria algoritmica della complessità*
With Newton da Costa and Francisco Antonio Doria:
Gödel's Way

Gregory Chaitin

PROVING DARWIN

Gregory Chaitin is widely known for his work on meta-mathematics and for his discovery of the celebrated Omega number, which proved the fundamental unknow-ability of math. He is the author of many books on mathematics, including *Meta Math! The Quest for Omega*. This is his first book on biology. Chaitin was for many years at the IBM Watson Research Center in New York. The research described in this book was carried out at the Federal University of Rio de Janeiro in Brazil, where Chaitin is now a professor. An Argentine-American, he is an honorary professor at the University of Buenos Aires and has an honorary doctorate from the National University of Cordoba, the oldest university in Argentina.

http://cs.umaine.edu/~chaitin

A Parable

Once upon a time a young rabbinical student went to hear three lectures by a famous rabbi. Afterwards he told his friends: "The first talk was brilliant, clear and simple. I understood every word. The second was even better, deep and subtle. I didn't understand much, but the rabbi understood all of it. The third was by far the finest, a great and unforgettable experience. I understood nothing and the rabbi didn't understand much either."

—Niels Bohr

From *Niels Bohr's Times* by **Abraham Pais**

PROVING DARWIN

Making Biology Mathematical

Gregory Chaitin

Vintage Books

A Division of Random House, Inc.

New York

FIRST VINTAGE BOOKS EDITION, FEBRUARY 2013

The illustrations at the beginning of each chapter are from
Kunstformen der Natur (Art Forms in Nature) by Ernst Haeckel.

The Library of Congress has cataloged the Pantheon edition as follows:
Chaitin, Gregory J.
Proving Darwin : making biology mathematical / Gregory Chaitin.
p. cm.
Includes bibliographical references and index.
1. Evolution (Biology)—Mathematics. 2. Biomathematics.
3. Evolution (Biology)—Mathematical models.
4. Biology—Mathematical models.
5. Evolution (Biology)—Philosophy. 6. Biology—Philosophy.
7. Computer Programming—Philosophy. I. Title.
QH371.3.M37C42 2012   576.8'2—dc23   2011040689

Vintage ISBN: 978-1-4000-7798-4

*Author photo © Buenos Aires Mathematics Festival
Book design by M. Kristen Bearse*

www.vintagebooks.com

146119709

Contradictory Quotes?

The chance that higher life forms might have emerged in this way [by Darwinian evolution] is comparable to the chance that a tornado sweeping through a junkyard might assemble a Boeing 747 from the materials therein.

> —**Fred Hoyle,**
> *The Intelligent Universe,* 1983

In my opinion, if Darwin's theory is as simple, fundamental and basic as its adherents believe, then there ought to be an equally fundamental mathematical theory about this, that expresses these ideas with the generality, precision and degree of abstractness that we are accustomed to demand in pure mathematics.

> —**Gregory Chaitin,**
> "Speculations on Biology, Information and Complexity,"
> *EATCS Bulletin,* February 2007

Mathematics is able to deal successfully only with the simplest of situations, more precisely, with a complex situation only to the extent that rare good fortune makes this complex situation hinge upon a few dominant simple factors. Beyond the well-traversed path, mathematics loses its bearings in a jungle of unnamed special functions and impenetrable combinatorial particularities. Thus, the mathematical technique can only reach far if it starts from a point close to the simple essentials of a problem which has simple essentials. That form of wisdom which is the opposite of single-mindedness, the ability to keep many threads in hand, to draw for an argument from many disparate sources, is quite foreign to mathematics.

—**Jacob T. Schwartz,**
"The Pernicious Influence of Mathematics on Science" (1960),
in *Discrete Thoughts: Essays on Mathematics, Science,
and Philosophy,* edited by Mark Kac, Gian-Carlo Rota
and Jacob T. Schwartz, 1992

Contents

Problem Solvers and Theorizers

by Gian-Carlo Rota[*]

Mathematicians can be subdivided into two types: problem solvers and theorizers. Most mathematicians are a mixture of the two although it is easy to find extreme examples of both types.

To the problem solver, the supreme achievement in mathematics is the solution to a problem that had been given up as hopeless. It matters little that the solution may be clumsy; all that counts is that it should be the first and that the proof be correct. Once the problem solver finds the solution, he will permanently lose interest in it, and will listen to new and simplified proofs with an air of condescension suffused with boredom.

The problem solver is a conservative at heart. For him, mathematics consists of a sequence of challenges to be met, an obstacle course of problems. The mathematical concepts required to state mathematical problems are tacitly assumed to be eternal and immutable.

Mathematical exposition is regarded as an inferior undertaking. New theories are viewed with deep suspicion,

[*]From *Indiscrete Thoughts* by Gian-Carlo Rota, Boston: Birkhäuser, 1997, pp. 45–46.

as intruders who must prove their worth by posing challenging problems before they can gain attention. The
problem solver resents generalizations, especially those
that may succeed in trivializing the solution of one of his
problems.

The problem solver is the role model for budding
young mathematicians. When we describe to the public
the conquests of mathematics, our shining heroes are the
problem solvers.

To the theorizer, the supreme achievement of mathematics is a theory that sheds sudden light on some incom-
prehensible phenomenon. Success in mathematics does
not lie in solving problems but in their trivialization. The
moment of glory comes with the discovery of a new theory that does not solve any of the old problems but ren-
ders them irrelevant.

The theorizer is a revolutionary at heart. Mathematical
concepts received from the past are regarded as imperfect instances of more general ones yet to be discovered.
Mathematical exposition is considered a more difficult
undertaking than mathematical research.

To the theorizer, the only mathematics that will survive are the definitions. Great definitions are what mathematics contributes to the world. Theorems are tolerated
as a necessary evil since they play a supporting role—or
rather, as the theorizer will reluctantly admit, an essential
role—in the understanding of definitions.

Theorizers often have trouble being recognized by the
community of mathematicians. Their consolation is the
certainty, which may or may not be borne out by history,

that their theories will survive long after the problems of the day have been forgotten.

If I were a space engineer looking for a mathematician to help me send a rocket into space, I would choose a problem solver. But if I were looking for a mathematician to give a good education to my child, I would unhesitatingly prefer a theorizer.

Dedicated to
John von Neumann (1903–1957)
Mathematician Extraordinaire

Preface

The purpose of this book is to lay bare the deep inner mathematical structure of biology, to show life's hidden mathematical core. This new field, which I call metabiology, is only three years old. Much remains to be done. It remains to be seen how relevant this theoretical work will be to real biology. Nevertheless, I feel that the time is ripe to present this new way of thinking about biology to the world.

The provocation for creating metabiology was my friend David Berlinski's delightfully polemical book *The Devil's Delusion,* which offers a stinging critique of Darwinism and a withering comparison of biological theory as contrasted with theoretical physics. This book is my answer to David; it is my attempt to find a remedy.

Proving Darwin is actually the course "Metabiology: Life as Evolving Software" that I am giving April–June 2011 at the Federal University of Rio de Janeiro in the wonderful program Epistemology and History of Science and Technology directed by my friend the poet/mathematician Ricardo Kubrusly. It is not a math course; it is more like a philosophy and history of ideas course on how and why to approach biology mathematically.

I hope you will enjoy reading *Proving Darwin* as much as I am enjoying teaching it. Teaching this course really made the ideas click in my mind, and everything finally fell into place.

Financial support for this research is being provided in Brazil by the Director of COPPE/UFRJ, Professor Luiz Pinguelli Rosa, and by the Brasilia CAPES Foreign Visiting Professor program.

I am grateful to the University of Buenos Aires and the Valparaiso Complex Systems Institute, which I visit frequently and where I lectured on these new ideas, which was very helpful. Many other institutions have also invited me to talk about metabiology: I am particularly grateful to Prof. Ilias Kotsireas for organizing a "Chaitin in Ontario" lecture series, to the University of Haifa for naming me a Caesarea Rothschild Institute Distinguished Lecturer, and to Jim Crutchfield and Jon Machta for inviting me to a meeting at the Santa Fe Institute, the "official debut" as it were of metabiology. The lecture that I gave in Santa Fe is Chapter 5 of *Proving Darwin.*

I am also grateful to Ana Bazzan and Silvio Dahmen for inviting me to visit the Federal University of Rio Grande do Sul. There I gave three talks on metabiology while simultaneously working on this book, which was very stimulating.

Furthermore, metabiology, which I define as a field parallel to biology and dealing with the random evolution of artificial software (computer programs) instead of natural software (DNA), would never have seen the light of day without my wife, Virginia Maria Fontes Gonçalves

Chaitin. It is very much a joint effort; our three-year-old child as it were. Virginia's field is philosophy.

However, this book is dedicated to John von Neumann, something rather unexpected. As I worked on the book, I began to feel more and more that I had him standing at my shoulder. You will see why.

Von Neumann was Hungarian but some people thought he was an extra-terrestrial only pretending to be human. However he was very smart, had studied human beings carefully and could imitate them fairly well!

The beautiful illustrations at the beginning of each chapter are all from *Kunstformen der Natur* (Art Forms in Nature) by Ernst Haeckel, and demonstrate the exuberant creativity of Nature, which this book attempts to provide a way of explaining. At the deepest level, seen from a vast distance, this is a corollary to Kurt Gödel's famous incompleteness theorem; it's a positive aspect of what may seem like an extremely negative theorem.

Biological creativity and mathematical creativity are not that different. Read this book and find out why!

PROVING DARWIN

Universidade Federal do Rio Grande do Sul,
Porto Alegre, Brazil, April 29, 2011
(Photo by Nicolas Maillard)

One

Introduction:
Overview of *Proving Darwin*

Like many pure mathematicians, I like giving "chalk" talks: improvised talks given on a blackboard or whiteboard using a minimum of technological assistance. Another strategy is for me to fill the board with what I want to cover just before I start to lecture, while people are still coming in, so that they can take in all the key ideas at a glance. In a large auditorium, however, a projector is necessary, or nobody will see anything.

On the facing page you can see me giving an overview of this book in a large auditorium at the Federal University of Rio Grande do Sul in southern Brazil. The four slides I prepared are on pages 5 and 6. They summarize Chapters 2 to 4, which outline my strategy for making biology mathematical. After you finish reading these chapters, you should review the slides. Then they will make more sense.

You've heard people refer to DNA as a computer program? Well, that's the whole idea: to make this metaphor into a mathematical theory of evolution. In fact, it turns out that the mathematical tools for doing this were already available in the 1970s. More precisely, we will treat evolution as a random walk in software space. Random walks are an idea that mathematicians feel comfortable with, although the space we are walking around in at random is bigger than usual.

I call this proposed new field "metabiology" because it is a highly simplified version of real biology—otherwise I wouldn't be able to prove any theorems. These theorems are presented in Chapter 5, which is the climax of the book, and was a talk that I gave at the Santa Fe Institute. Chapters 6 to 8 discuss the broader significance of metabiology, theological, political and epistemological. And then there are two appendices.

In the first appendix you can read the crucial section on self-reproducing automata in John von Neumann's far-seeing "DNA = Software" paper that influenced Sydney Brenner, who in turn influenced Francis Crick—a remarkable fact that I only discovered while working on this book. And the second appendix gives some additional mathematical details that may be of interest to experts.

Finally, I give a short list of suggested further reading, some books and a few articles that are important if you want to really understand metabiology. These are the books and articles that helped me the most to come up with a strategy for making biology mathematical, plus a few more related items that I threw in just for the fun of it. Enjoy them all!

LIFE AS EVOLVING SOFTWARE
Artificial Digital Software:
  Computer Programming Languages, 50/60 years old
Natural Digital Software:
  DNA, $3-4 \times 10^9$ years old
DNA = Universal Programming Language
  Life = Evolving Software
  Biology = Software Archeology (evo-devo!)
  Origin of Life = Origin of Software

  Biological Creativity = Math Creativity
  Gödel Incompleteness $\rightarrow$ Unending Evolution

THE HUMAN DISCOVERY OF SOFTWARE
History of Molecular Biology
  Schrödinger, *What Is Life?*
  Discovery of Software: Turing/von Neumann, 1936/1951
  Alan Turing $\rightarrow$ John von Neumann $\rightarrow$ Sydney Brenner $\rightarrow$
    Francis Crick
History of Metabiology
  Definition of life as something that evolves (John Maynard
    Smith, 1986)
  Mathematical proof there is something satisfying the
    definition (2010)
Uses Postmodern (Post-Gödel) Mathematics
  Algorithmic Information Theory, Computability Theory
  Complexity Theory, Computer Science

THE MATHEMATICS OF METABIOLOGY (PART 1)

Our Toy Model of Evolution

   Single Mutating Software Organism Calculates Single Integer
      & Halts

   Fitness of Organism = Integer It Calculates

   Requires Creativity: $N \rightarrow N + N \rightarrow N \times N \rightarrow N^N \rightarrow N^{N^{N^{\cdots}}}$ N times
      etc.

   Evolution = Hill-Climbing Random Walk in Software Space
      (Increasing Fitness)

   Try K-bit algorithmic mutation M from A to $A' = M(A)$ with
      probability 2^{-K}

   Mutation M works only if $A' = M(A)$ is fitter than original organ-
      ism A

   Need oracle to eliminate mutations that give no A' or an A' that
      doesn't halt

   Mutation Distance between A and B = $-\log_2$ (probability to go
      from A to B in a single mutation)

      = size in bits of smallest program M with $B = M(A)$

THE MATHEMATICS OF METABIOLOGY (PART 2)

To measure speed/rate of biological creativity we use

   BB(N) = Busy Beaver function of N

      = largest fitness of any $\leq$ N bit program

   Calculating BB(N) requires N bits of inspiration

   BB(N) grows faster than any computable function

Different Evolution Regimes

   "Exhaustive Search" reaches fitness BB(N) in time 2^N

   "Intelligent Design" reaches fitness BB(N) in time N

   Random Evolution reaches fitness BB(N) in time between
      N^2 and N^3

Nota Bene

   If organisms are improved mechanically, algorithmically,
      i.e., the sequence A, A', A'' . . . is computable,
      then fitness can only grow as a computable function

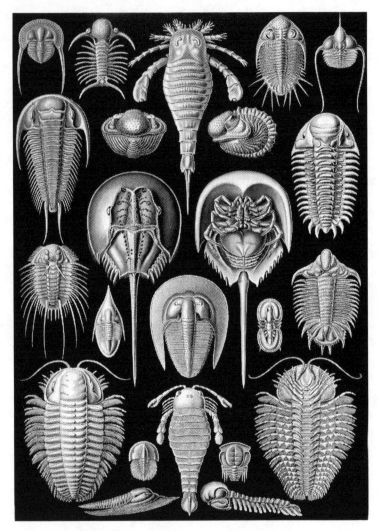

A page from Ernst Haeckel's *Kunstformen der Natur*
(Art Forms in Nature).

Two

Life as Evolving Software:
The Evolution of Mutating Software

Look out the window on a summer's day, and what do you see? Bushes, trees, flowers, birds, insects . . . a tremendous variety of living beings. Nature's diversity and richness are quite overwhelming . . . Can all this exuberant creativity be explained by Darwin's theory of evolution? Biologists are convinced that this is so. But if you compare theory in biology with theory in physics, and if you look at biology as a mathematician, things are not that convincing. There is empirical evidence in favor of Darwin's theory, but there is no mathematical proof.

Remember that Institute for Advanced Study dinner where a young astrophysicist proudly described his latest discovery to Kurt Gödel, only to be told "I don't believe in empirical science; I only believe in *a priori* truths!" Well, Gödel had a point. If Darwin's theory is as fundamental as biologists think, then there ought to be a general, abstract mathematical theory of evolution that captures the essence of Darwin's theory and develops it mathematically.

And that is what we set out to do here. Remember Kepler's *Harmonices Mundi*, Newton's *Philosophiae Naturalis Principia Mathematica*, and Laplace's *Exposition du système du monde*, that first revealed the mathematical structure underpinning the physical world? (I am happy to have a two-hundred-year-old copy of Laplace's book.) How can we do the same for biology, a very, very different kind of science from physics?

Well, not by using the differential equations of theoretical physics! To develop a theoretical physics for biology, a fundamental mathematical theory for biology, we must use a different kind of mathematics. Differential equations will not do, not at all.

What is biology really about? Well, it is about information. Even theoretical physics, even quantum mechanics, is now starting to be about information: qubits. But biology is about a different kind of information: algorithmic information. When people say DNA is like a computer program, when evo-devo (evolutionary developmental biology) describes the DNA program for the development of an embryo, that is algorithmic information. And that is the kind of information we must build into a new kind of theoretical physics, one that applies to biological systems, to living beings.

Even after Kepler and Newton and Laplace, some people thought that biology was different, that living beings have a divine spark. Well, according to Darwin, we do not have a divine spark. In effect the whole world is divine, creative. Life appears by itself by chance, not by design. Here we propose a mathematical theory for this, which I call "metabiology."

Metabiology is a new kind of theoretical physics. Metabiology is further from real biology than theoretical physics, using differential equations, is from physics, because the mathematical structure of biology is harder to see. On the face of it, biology is too complicated and has too many exceptions for there to be a deeply mathematical theoretical biology that is like theoretical physics, where the physical world is built, so to speak, directly out of math.

Metabiology deals with software; that is what genetic information is, that is what DNA is. Take a good look at Neil Shubin's wonderful book on evo-devo, *Your Inner Fish.* Our bodies are full of software, extremely ancient software. We have subroutines from sponges, subroutines from amphibia, subroutines from fish. There is a stage in gestation in which the human embryo has gills! Each cell contains a complete copy of this DNA software, and in effect contains the entire history of the organism, because evolution makes minimal changes, it tries to change as little as possible, just like what happens with large human software projects. You can't start over, you have to make do with what you have as best you can. As Jacques Monod said, Nature is a *bricoleur,* a handyman, a tinkerer. You make do with old things, you patch them up, you fix them so you can reuse them.

In fact, it's like archeology—that's what biology really is, a kind of software archeology! So there is artificial software, computer programs, and there is natural software, DNA. Nature invented software before we did, long before. And the origin of life is really the origin of software, the origin of DNA, a universal programming language found in every cell. A powerful programming language, one that can presumably express any possible

algorithm, any set of instructions for building and running an organism. It is a programming language that we
are beginning to understand, a very complicated programming language, one that has grown by accretion over the
millennia, like those Hindu temples carved out of solid
rock with Gods on top of Gods on top of Gods . . . Like
wine bottles used for many years as candlestick holders
without the melted wax being removed . . .

Our artificial software, our programming languages,
are much simpler than DNA, and we know how they
work, because we designed them and they are only half a
century old not billions of years old. So instead of studying randomly evolving natural software, DNA, we will
develop a parallel theory, a theory of randomly evolving artificial software, randomly evolving computer programs. That's what metabiology is about. That's much
simpler than real evolution, hopefully simple enough that
we can prove theorems about it. Hopefully simple enough
that we can understand precisely what is going on, precisely how it works.

In fact, the notion that the world is built out of mathematics does not start with Kepler and Newton and
Laplace, it starts with Pythagoras. And metabiology is a
kind of Pythagorean biology. The ancient Greeks started
with a world, with a mythology, in which everything is
alive: the capricious Gods, the Sun, the Wind, the Rivers,
the Trees . . . Later the Greeks switched to *logos,* to the
belief that the universe is governed by laws.

To Pythagoras, not only the laws of Nature are mathematical, the fundamental ontology of the world is math,

the world is built out of mathematics. Modern theoretical physics follows Pythagoras. To Plato, the world of (mathematical) ideas is more real than the real world. The world of ideas is static, eternal, perfect; the real world, the world of appearences, is ephemeral. But Life is plastic, creative! How can we build this out of static, eternal, perfect mathematics?

The same tension continues during the Enlightenment before the French revolution, which was against religion as much as it was against the monarchy. Denying that life contains a divine spark, La Mettrie penned his famous little book *L'Homme machine* (Man a Machine, Machine Man) (1748). La Mettrie was a doctor who cut people open and saw how they functioned: a very complicated machine, but only a machine.

We have gone a little further than La Mettrie. Now we know all about computers and about the distinction between hardware and software. Yes, human beings are machines, but if you want to understand evolution, you have to concentrate on the software, which is what evolves and changes the hardware. The software is more important than the hardware. So that's what this book is about: *L'Homme software* not *L'Homme machine*.

In French films from the 1930s by Marcel Pagnol, the village priest and the village schoolteacher would be the best of friends even though one was a believer and the other was an atheist. They would always be kidding each other about this. And the same tension continues in the USA today, in the political battles between Creationists and more conventional biologists.

Even now, conventional biologists are surprised that single-celled life without nuclei appeared on Earth after only 200 million years, while cells with nuclei took 2 billion years to develop, and suspect that life on Earth was either seeded by accident, which is called *panspermia,* or deliberately planted, so-called *directed panspermia.* In particular, you can find this in Francis Crick's book *Life Itself,* in Fred Hoyle's *The Intelligent Universe,* and in Martin Nowak's *Supercooperators,* so it is definitely not an extreme minority view. The prescient book by Hoyle even compares biology to a very messed up computer program that he evolved by accretion to simulate supernovas—very much a central theme of metabiology and of evo-devo as explained in Shubin's *Your Inner Fish.*

To repeat: Life is plastic, creative! How can we build this out of static, eternal, perfect mathematics? We shall use postmodern math, the mathematics that comes after Gödel, 1931, and Turing, 1936, open not closed math, the math of creativity in fact.

We need an open, non-reductionist kind of math because the creativity of the biosphere is a key issue. **Biological creativity**—biological inventiveness and richness and diversity—has somehow gotten lost in the standard accounts of evolution. But this was not at all the case in the work of Ernst Haeckel, the Darwin of the German-speaking world. Haeckel became wealthy presenting his own version of Darwin's theory to the German-speaking world. His books were best sellers, and his immense home is now a biology museum. One of the pieces of evidence that Haeckel gave for evolution was

Phyletisches Museum

his famous (or infamous) doctrine that *ontogeny recapitulates phylogeny,* which means that during development an embryo more or less repeats the entire evolutionary history of that organism, a theme now taken up in expanded and corrected form by evo-devo.

Above is a photo taken by my wife, Virginia, of Haeckel's home in Jena that is now a museum. Note the tree of life on the wall of the house and the words "ontogenie" and "phylogenie"!

Anyone interested in biological creativity should see these two books by Haeckel containing marvelous drawings of an amazing diversity of life-forms: *Art Forms from the Ocean* and *Art Forms in Nature* (*Kunstformen der Natur*). They were reprinted by Prestel in 2009 and 2010 with added historical commentary.

See also Stephen Gould's *Wonderful Life* on the Cambrian explosion and a plethora of amazing body plans that were briefly tried by Nature, a combinatorial explosion exploring all simple possibilities, all simple programs, in fact. This is what Stephen Wolfram terms "mining the computational universe," trying all possible simple programs for something, which he frequently exploits as a design strategy—see his book *A New Kind of Science.*

So Nature is tremendously creative, tremendously inventive. Let's consider a popular mathematical approach to biology, population genetics, which was developed by Fisher, Wright, Haldane, Hamilton, Maynard Smith, Dawkins, Nowak and others, a beautiful field. Unfortunately population genetics **defines** evolution to be changes in gene frequencies in a population because of competition or selective pressures due to the environment. The finite gene pool is fixed, and therefore there is **no** creativity. In contrast metabiology works with a very rich space of possibilities, a software space, which gives us a way to talk about where new genes come from. That's the whole point of metabiology, in fact. But metabiology does not pay much attention to populations, competition or the environment. So population genetics and metabiology are complementary; they deal with different issues.

To repeat once more: Life is plastic, creative! How can we build this out of static, eternal, perfect mathematics? Here is the answer: Life is creative, plastic software; physics is rigid, mechanical hardware!

The biosphere is full of software, every cell is run by software, 3- to 4-billion-year-old software. Our artificial software is only fifty years or so old. *But we could not realize that the natural world is full of software, we could not see this, until we invented human computer programming languages.* The world was full of software even before we knew what that was! Software is the reason for the plasticity of the biosphere—normal machines are rigid, mechanical, dead. Software is alive!

Furthermore *the origin of life is the origin of software,* the spontaneous appearance of entities governed by software (cells), and of the DNA language for this software. Every cell in our body has the complete DNA for an entire human being, even though only some of this software is used depending on the tissue. And every living organism on Earth uses essentially the same DNA language—so far, there is no evidence of independent creations of life, independent origins of life.

Nobel laureate Sydney Brenner shared an office with Francis Crick of Watson and Crick. Most molecular biologists of his generation credit Schrödinger's book *What Is Life?* (Cambridge University Press, 1944) for inspiring them. In his autobiography *My Life in Science* Brenner instead credits von Neumann's work on self-reproducing automata. Von Neumann's Hixon Symposium paper "The General and Logical Theory of Automata" that inspired Brenner contains many essential biological ideas, developed even **before** Watson and Crick discovered DNA, in Brenner's opinion a truly remarkable mathematical anticipation.

Many years ago I had the privilege of meeting Brenner at an MIT meeting on the physics of computation that was organized by Rolf Landauer and Edward Fredkin. Brenner gave a marvelous talk on molecular biology and was interested to see how automata theory continued developing after von Neumann.

As I learned by reading von Neumann's "The General and Logical Theory of Automata" as a student, it was Turing's 1936 paper "On Computable Numbers, with an Application to the *Entscheidungsproblem*" that created the idea of flexible machines, **universal** machines, the general-purpose computer, and the distinction between software and hardware. DNA is presumably a universal programming language, one that is sufficiently powerful to express any algorithm. Brenner credits von Neumann with the idea that DNA contains the software for building (and running) the organism, an idea that is now commonplace due to **evo-devo**, evolutionary developmental biology, which studies how the formation of embryos evolves.

As I said before, this is very visible in Shubin's book, which shows how we have subroutines from sponges, fishes and amphibians. To understand some strange things in the human body, some strange features in its design, you have to compare that with how fish are built. Minimal changes were made to convert fish into mammals!

Let's return now to the question Erwin Schrödinger asked in 1944: "What is life?" In his 1986 Oxford University Press book *The Problems of Biology* John Maynard Smith—whom I met at a meeting north of the Arctic Circle in Abisko, Sweden—has a chapter discussing this. Flames have a metabolism—they take in material and expel material while retaining their form—and furthermore they self-reproduce, but they do not have heredity and mutations or evolve. According to Maynard Smith, life is what evolves, what is plastic, creative.

My Pythagorean biology is a mathematical proof that life exists. That is, I construct artificial mathematical life-forms. I have a minimal model of biology, a toy model, all the essential concepts and features, because I can prove it evolves by Darwinian evolution. I have found an evolving life-form in the Pythagorean world of pure mathematics!

My organisms have no metabolism, no bodies, only DNA; no hardware, only software. I study the evolution of mutating software, in fact a single mutating organism, which gives me a random walk in software space. The space of all possible computer programs is a space rich enough to model all possible designs for an organism.

Is this too simple? Not if I can get life to evolve! Physicists are very comfortable with what they admiringly call "toy models." Picasso said that "Art is a lie that helps us to see the truth." Similarly, theories are lies that help us to see the truth. But biologists think that every detail counts; they do not distinguish between what is fundamental and what is secondary. After all, they worked hard to discover all that stuff. However as Brenner emphasizes in his autobiography—and this is quoted in James Gleick's *The Information*—the energetics, the metabolism of living organisms is unimportant, all that counts is the information, all that counts is where you get the instructions for doing something. The energy will take care of itself!

So our DNA is a very, very old piece of software,

tremendously patched, not at all clean and elegant and well designed. If we could start over and directly design mammals, we could do much better. But we cannot start over, and neither can the huge world of artificial software, of software technology, that was created by humans and that is only half a century old instead of billions of years old.

In particular, we and chimpanzees have almost the same genes that code for proteins. But a lot of the DNA used to control which genes are expressed is different. In other words, the low-level subroutines do not change much, because many people depend on them, they are used too much. But the higher-level, newer software can change much more easily.

In just fifty years the human programming environment has become extremely high-level: graphics interfaces, ways to get material from the Internet. Nobody today would want to start over and program the bare metal in assembly language like I originally did to earn a living. We take our extremely high-level programming languages and programming environments for granted. We cannot begin again. Decisions are frozen into our current technology, like the inefficient "qwerty" typewriter keyboard designed to keep early typewriters from jamming if people typed too fast, a problem modern keyboards do not have.

The same way that von Neumann's mathematics anticipated biological discoveries that were made afterwards, my work, metabiology, for the math to be beautiful requires **algorithmic** mutations, not point mutations, high-level mutations, not low-level mutations. It is not clear to what extent algorithmic mutations occur in biology. So, as my wife, Virginia, pointed out to me, metabiology raises the issue of how high-level the mutational mechanisms are in actual organisms.

That concludes this chapter. By now you should have a fairly good idea of our overall life-as-evolving-software approach. We've seen that Nature discovered software a long time ago. In the next chapter we'll study in some detail the relatively recent human discovery of software, which has a surprising history. And in the chapter after that I'll explain how to take the idea of life as evolving software and make it into a mathematical theory where you can prove things. Then we'll be ready for Chapter 5, my Santa Fe Institute lecture, the climax of this book.

Caveat: Metabiology in its present form cannot address thinking and consciousness, fascinating though these be.

Kunst-Formen der Natur

von

ERNST HAECKEL

Leipzig und Wien

Bibliographisches Institut

Three

The Human Discovery of Software:
Turing and von Neumann as Biologists

In this chapter we present a revisionist history of the discovery of software and of the early days of molecular biology, as seen from the vantage point of metabiology; the present is always rewriting the past in order to justify itself. As Jorge Luis Borges points out, one creates one's predecessors!

As an example of this history rewriting process, consider the atheist modern scientist mechanical worldview image of Newton created by Voltaire. In his essay "Newton, the Man," John Maynard Keynes describes Newton as "the last of the magicians, the last of the Babylonians and Sumerians," not at all the first modern scientist, and closer to Doctor Faustus than to Copernicus. Newton spent more of his time on alchemy and theology than on mathematics and physics; he assembled a remarkable collection of medieval alchemy books. My friend Stephen Wolfram has three-hundred-year-old books on theology by Newton and Leibniz (the originals, not copies) next to each other on a shelf; Newton's *Principia* is unaffordable, but not that many people are interested in theology.

In fact the past is not only occasionally rewritten, it *has to be* rewritten in order to remain comprehensible to the present. On the topic of how one should write the history of science, see Helge Kragh's brilliant *An Introduction to the Historiography of Science.*

Let's now re-examine the early history of computing theory and molecular biology in the light of metabiology, and see the threads that led—or that should have led, if scientific progress were entirely rational—to metabiology.

Our story is full of surprises, starting with questions about philosophy and the foundations of mathematics, and including the creation of one trillion-dollar technology, which has already happened—and this may well soon be followed by the creation of a second such game-changing technology. There are even connections with medieval magic.

Would you like to know more? Read on!

Few people remember Turing's work on pattern formation in biology (morphogenesis), but Turing's famous 1936 paper "On Computable Numbers" exerted an immense influence on the birth of molecular biology indirectly, through the work of John von Neumann on self-reproducing automata, which influenced Sydney Brenner who in turn influenced Francis Crick, the Crick of Watson and Crick, the discoverers of the molecular structure of DNA. Furthermore, von Neumann's application of Turing's ideas to biology is beautifully supported by recent work on evo-devo (evolutionary developmental biology). The crucial idea: DNA is multibillion-year-old software, but we could not recognize it as such before Turing's 1936 paper, which according to von Neumann creates the idea of computer hardware and software.

We discussed this crucial idea in the previous chapter; perhaps before proceeding we should summarize what was covered there:

Hardware	Physics	Dead	Rigid	Closed	Mechanical
Software	Biology	Alive	Plastic	Open	Creative

Natural Software	DNA	$3\text{-}4 \times 10^9$ years old
Artificial Software	Computer Programs	50 years old

Newtonian Math	Continuous Math, Differential Equations	For Physics
Postmodern Math	Discrete, Combinatorial, Algorithmic	For Biology

Definition of life (John Maynard Smith, *The Problems of Biology,* 1986)
Mathematical proof that something exists satisfying the definition (2010)

Remember Molière's *bourgeois gentilhomme* who discovered to his amazement that all his life he had been speaking prose? We are that gentleman. Our bodies are full of software, they always have been, but before we could recognize natural DNA software as such, we had to invent artificial software, human computer programming languages.

There was software all around us, in every cell, ancient software, but we couldn't see that until we invented software ourselves! Furthermore as evo-devo shows, organisms contain their own history, as per Shubin's *Your Inner Fish,* your inner sponge, your inner amphibian . . . Biology is just a strange kind of archeology, software archeology.

And the purpose of human love-making is to integrate software from the male (in sperm) with software from the female (in the egg). That is why people fall in love, because they want to combine their subroutines. So, in a sense, von Neumann discovered why people fall in love.

Furthermore, the origin of life, which is still deeply mysterious, is the origin of software—natural software, not artificial software. But help is on the way. Stephen Wolfram's *A New Kind of Science* can be reinterpreted as a book about the origin of life. One of Stephen's main points is that it is very easy to get a combinatorial symbolic system to be a universal Turing machine or a general-purpose computer. It is very easy to build a computer out of almost any discrete math components. He refers to this, I believe, as *the ubiquity of universality.* At a philosophical level, then, this means that the origin of life is not in general that surprising, but perhaps in the particular implementation here on Earth it is. Anyway, thank you, Stephen!

Enough generalities! Now let me tell you in more detail how the computer was invented by Alan Turing (and also simultaneously by Emil Post) to help clarify a question about the foundations of mathematics—Nature, which invented the computer and hardware/software first, doesn't care a fig about the foundations of math, but it does care about plasticity.

First let's talk about an old dream, certain knowledge. Or, as the amazing polymath Leibniz put it, mechanical knowledge, reasoning as certain as arithmetic, truths as obvious as $2 + 2 = 4$. No disputes anymore, said Leibniz: Gentlemen, let us compute and determine who is correct! What a beautiful dream!

Leibniz did not do too much work developing what today is called symbolic logic or mathematical logic, but he stated the goal extremely clearly and forcefully, and for this reason he is considered the father, or the grand-father, of modern logic. Leibniz could not devote too much time to any one topic; he was omnivorous, he was interested in everything. For example, he also invented binary arithmetic, and calculating machines that could multiply—Pascal's original calculating machine, the *Pascaline,* could only add and subtract.

Through the years many logicians worked on Leibniz's dream of certain knowledge, of mechanical reasoning: de Morgan, Boole, Peano, Frege, Russell, Hilbert, Gödel, Turing, Post . . . But it didn't work, it turned out it **could not work.** Instead of certain, mechanical knowledge, Gödel found incompleteness and Turing found uncomputability. But in the process Turing also found complete/ universal programming languages, hardware, software, and universal machines.

One milestone was the German mathematician David Hilbert's improved version of Leibniz's dream. Hilbert wanted a formal axiomatic theory for all of mathematics, a mechanized version of Euclid's *Elements* that would cover all of math, not just geometry. The key point was that it should be possible to check mechanically whether or not a proof is correct, whether or not the reasoning follows all the rules. To do this one would need to invent a meticulous artificial language sufficiently powerful to express *all possible* mathematical reasoning, all possible mathematical proofs.

In 1931, Gödel showed this is impossible: no such mechanical universal language for reasoning can ever be found, can ever enable us to prove all mathematical truths. This is called *incompleteness.* But then in 1936 Turing showed that there are in fact complete or universal mechanical languages for performing mathematical calculations instead of expressing mathematical proofs. And the rest is history: the modern computer was born!

How did Gödel refute Hilbert and Leibniz? By constructing an arithmetical mathematical assertion that asserts its own unprovablity: "I am unprovable," which is provable if and only if it is false.

Turing's technique was different, less tricky, deeper. He studied what machines could compute, and observed that most real numbers are uncomputable and therefore have numerical values that cannot be determined via formal proof, for otherwise one could mechanically run through all possible proofs to systematically calculate the value of these uncomputable real numbers.

This is actually my preferred version of Turing's basic result. The usual way it is explained is in terms of the famous *halting problem.* Turing shows that there is no systematic way, no mechanical procedure, no formal axiomatic theory, for deciding whether or not a self-contained computer program will eventually halt. You can start it running and calculate step by step, but to decide if it goes on forever or not is in the general case quite impossible.

So we get computer programming languages, universal ones, ones that are powerful enough to write *any algorithm.* But we lose certainty, we lose mechanical reasoning. That dream is gone forever.

Not to worry! According to Emil Post—who is not as well-known as Gödel and Turing but was at their level (he came up with Turing machines too, and also with an incompleteness theorem that remained unpublished for years)—the axiomatic method, and especially Hilbert's formal axiomatics, was just a terrible mistake, just a confused misunderstanding.

According to Post, math **cannot** provide certainty because it is not closed, mechanical, it is creative, plastic, open! Sound familiar? You bet, we have been talking about biological creativity all through the previous chapter, and now we find something like it in pure math too! So math is creative, not mechanical, math is biological, not a machine! I told you that mathematical and biological creativity are not that different—we'll see that in even more detail in the next chapter.

The point is particularly well made in the titles of two of philosopher Paul Feyerabend's books: *Against Method* and *Farewell to Reason.* Feyerabend advocates creativity and imagination—in a word, anarchy—in science based on his reading of the history of science and without ever mentioning Gödel or Turing. But in my opinion *Against Method* would be the best title for a book on the unsolvability of Turing's halting problem, and *Farewell to Reason* would be the best title for a book on Gödel's incompleteness theorem. What Feyerabend believes to be the case in the world of science for purely philosophical reasons, in the world of mathematics are actually mathematical theorems—there provably are no general methods for solving all mathematical problems.

Nevertheless, as the mathematican Gian-Carlo Rota has observed in his essay "The Pernicious Influence of Mathematics upon Philosophy," philosophy is actually the art of finding bad reasons for what one believes instinctively, somewhere deep in the gut. Because of subconscious childhood emotional cravings, in fact. Rota made few friends in the philosophy community with clever remarks like this, but I love philosophy and I also think that Rota has a point: Philosophy should not try to imitate mathematics too much, especially not the formal axiomatic method that Hilbert championed. For as Rota observes, if we could define our terms precisely, that would be the end of philosophy. Formal axiomatics is not creative . . .

There are, in fact, different kinds of mathematical creativity. Some people, the majority, are interested in creativity within a formal mathematical axiomatic theory = find a proof but stay within the current paradigm = normal science (Kuhn). I myself am more interested in "savage creativity" (Deleuze) = changing the formal theory = new axioms, new concepts = paradigm shift (Kuhn) = against method (Feyerabend)!

And now for von Neumann's self-reproducing automata. Von Neumann, 1951, takes from Gödel the idea of having a description of the organism within the organism = instructions for constructing the organism = hereditary information = digital software = DNA. First you follow the instructions in the DNA to build a new copy of the organism, then you copy the DNA and insert it in the new organism, then you start the new organism running. No infinite regress, no homunculus in the sperm!

For the details, see the crucial part of von Neumann's paper on self-reproducing automata in the first appendix. This was the paper that inspired Sydney Brenner to go to Cambridge to work with Francis Crick. Crick needed someone to work with him, for Watson had returned to the States after publishing their famous paper in *Nature*.

First Crick shared an office with Watson, then with Brenner; he could not work alone, he needed to kick ideas around, he needed to have someone to talk to all day long. Francis Crick was the supreme theoretician, the strategist, the person who masterminded the creation of molecular biology. And half of Crick was Brenner. You see, after Watson and Crick discovered the molecular structure of DNA, the 4-base alphabet A, C, G, T, it still was not clear what was written in this 4-symbol alphabet, it was not clear how DNA worked. But Crick, following Brenner and von Neumann, somewhere in the back of his mind had the idea of DNA as instructions, as software. And he knew that what counted, what was important, was how this information flowed around the cell, and how it turned into protein . . .

How did Brenner hear about von Neumann's Hixon

Symposium paper? Back in South Africa, before he went to Oxford and then to Cambridge, Brenner had as classmate the computer scientist Seymour Papert, who later worked with Marvin Minsky at MIT. And it was Papert who alerted Brenner to von Neumann's paper. (I never met Papert, but I do know Marvin.)

Von Neumann's work on self-reproduction didn't stop in 1951 with his so-called *kinematical model,* essentially the plan for a physical device. Following a suggestion by Stanislaw Ulam (whom I had the privilege of meeting at Los Alamos), von Neumann's model of self-reproduction then moved from the physical world to a toy graph paper two-dimensional plane divided into identical squares, each a finite-state machine, a so-called cellular automata (CA). This was published posthumously.

The cellular automata world used by von Neumann is a homogeneous, uniform, totally plastic world, a world where everything is software, information, not hardware, a world where magic applies: the right magic spell will create anything. In it there is a universal constructor as well as a universal computer. That sounds good, but in this CA world there are some problems too: it is easier for an organism to reproduce itself than it is for it to move (translate itself).

It has taken longer than it did with Turing's work on the universal machine, but von Neumann's work on self-reproduction and universal constructors is starting to have technological applications: namely printers for objects, three-dimensional printers, new flexible manufacturing technology, 3D printers that can print themselves (http://reprap.org), ultimately perhaps, the universal factory that can build anything!

So you see, Turing's paper has had a tremendous impact, and may have more. The computer is not just a tremendously useful technology, it is a revolutionary new kind of mathematics with profound philosophical consequences. It reveals a new world.

I have devoted most of my life to exploring one little aspect of that new world: using the size of computer programs as a complexity measure, defining randomness as irreducible complexity, as algorithmic incompressibility, and studying a number I call the halting probability Ω. Ω compactly tells us about individual instances of Turing's halting problem. If we could know the numerical value of Ω with N bits of precision, that would enable us to answer the halting problem for all programs up to N bits in size. Ω is jam-packed with logically and computationally irreducible mathematical information.

The halting probability of a program generated by coin tossing is a paradoxical real number: In spite of Ω's simple definition, its numerical value is maximally uncomputable, maximally unknowable, and shows that pure mathematics contains infinite irreducible complexity. Ω can be interpreted pessimistically, as indicating there are limits to human knowledge. The optimistic interpretation, which I prefer, is that Ω shows that one cannot do mathematics mechanically and that intuition and creativity are essential. Indeed, in a sense Ω is the crystallized, concentrated essence of mathematical creativity—Emil Post all over again.

Furthermore, as I stated for the first time in an article in the *Bulletin of the European Association for Theoretical Computer Science* (*EATCS*) in February 2007, the infinite irreducible complexity of the halting probability Ω shows that pure math is even more biological than biology, which is very complicated but only has finite complexity. Pure math has infinite complexity. It was this clue that led me to try to create metabiology, which I announced in the *EATCS Bulletin* in February 2009, only two years later.

So you see, Gödel, Turing, Post and von Neumann opened a door from math to biology; they gave us the necessary conceptual tool-kit. We need postmodern discrete algorithmic math to understand biology, not Newtonian differential equations, not old math, not analysis.

On pages 35–37 is a timeline and some references to the work we have discussed in this chapter. In the next chapter we'll get down to the nitty-gritty and see how to build a mathematical theory about randomly mutating software. Following John Maynard Smith's and Sydney Brenner's advice, we shall ignore bodies and metabolism and energy and consider purely software organisms. I'll tell you how I go about building a mathematical theory, which I've succeeded in doing once before (algorithmic information theory), and hopefully now once again (metabiology). And as you will see, the criteria for success are mostly aesthetic; math is an art form.

We'll be able to prove that Darwinian evolution works in our toy model, and amazingly enough, the organisms that evolve by natural selection are better and better lower bounds on the halting probability Ω, which I did not at all foresee. A pleasant surprise indeed. But in retrospect inevitable, not surprising, as we will see in Chapter 5.

Kurt Gödel, 1931*	**Self-reference**: "This statement is unprovable!"	**Incompleteness** of formal systems for mathematical reasoning	
Alan Turing, 1936*	On computable numbers, with an application to the *Entscheidungs-problem*	**Completeness** of formal systems for mathematical computations = Universal Programming Languages, Universal Turing Machines = general-purpose computers	Theoretical Philosophy of Math Paper creates idea of software, trillion-dollar computer industry [von Neumann]!
Turing's late work on biology	Morphogenesis = Newtonian Math = Partial Differential Equations = Obsolete!	**It was von Neumann who saw how Turing's work applied to biology, not Turing himself!**	**Von Neumann starts mathematical biology! Beginning of metabiology**

*Martin Davis, *The Undecidable: Basic Papers on Undecidable Propositions, Unsolvable Problems and Computable Functions,* Dover, 2004.

John von Neumann, late 1940s, early 1950s	Theory of self-reproducing automata (**self-reference** becomes **self-repro-duction!**)**	Universal Constructors, Printers for Objects, 3D Printers, Flexible Manufacturing, Self-Reproducing Printers!	Universal Factory = Future Trillion-Dollar Business?!
Von Neumann Died Young (53)	Died 1957; Watson, Crick Paper on DNA = 1953		
Sydney Brenner, late 1950s, 1960s	Biologist inspired by von Neumann, not by Schrödinger's *What Is Life?***	Influenced Crick, Shared office, All that matters is information = instructions for building things, doing things = discrete algorithmic software	Nobel Prize winner

John von Neumann, "The General and Logical Theory of Automata," delivered 1948 at the Hixon Symposium on Cerebral Mechanisms in Behavior, Pasadena, California, and published in Lloyd Jeffress, *Cerebral Mechanisms in Behavior: The Hixon Symposium,* John Wiley and Sons, New York, 1951, pp. 1–41; John Kemeny, "Man Viewed as a Machine," *Scientific American* **192 (April 1955), pp. 58–67 (describes von Neumann's self-reproducing automata); E. F. Moore, "Artificial Living Plants," *Scientific American* **195** (October 1956), pp. 118–126 (another reaction to von Neumann); John von Neumann, *Theory of Self-Reproducing Automata,* University of Illinois Press, Urbana, 1966 (posthumous; edited and completed by A. W. Burks).

***Sydney Brenner, *My Life in Science,* Biomed Central Ltd., 2001; Matt Ridley, *Francis Crick,* Eminent Lives, 2006.

Stanislaw Ulam****	Cellular Automata World, 29 states, 4 neighbors	Totally Plastic World, Everything is Software, Magic Incantations, Information!	World as Idea! Down with Materialism!

****Stanislaw Ulam, *Adventures of a Mathematician,* 2nd edition, University of California Press, Berkeley, 1991 (1st edition, 1983). See also Konrad Zuse, *Rechnender Raum* (Calculating Space), Friedrich Vieweg & Sohn, Braunschweig, 1969. An English translation of *Rechnender Raum* will appear in Hector Zenil, *A Computable Universe,* World Scientific, Singapore, 2012.

On the plasticity of the world see Freeman Dyson's vision of a totally green technology *The Sun, the Genome, and the Internet,* Oxford University Press, 2000: seeds that grow into houses instead of trees, children performing genetic engineering to design new flowers, etc. See also Allison Coudert, *Leibniz and the Kabbalah,* and Umberto Eco, *The Search for the Perfect Language,* on the Adamic language used by God to create the world and whose structure directly reflects the fundamental inner structure of the world. Knowledge of this language would give us God-like powers (as in Jorge Luis Borges' tale "The Rose of Paracelsus").

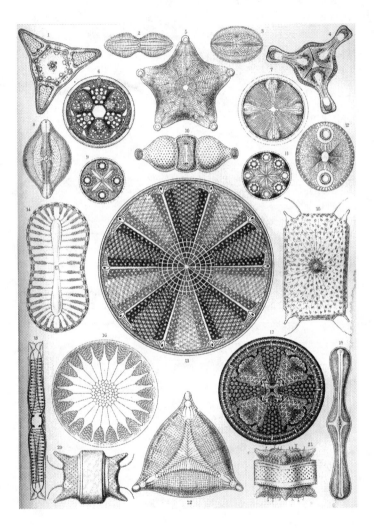

Four

The Mathematics of Metabiology: Random Walks in Software Space

As we have seen in the previous two chapters, life as evolving software is a nice metaphor. But is it a mathematically fertile approach? That is the key question. I think that it is, since using it we can prove that evolution works in a fairly straightforward and natural way, which I will now explain . . .

But before leaping into the mathematical formulation of metabiology, I want to set the stage. I want to explain how pure math works, how it functions, what you can expect it to do for you, so that you will look upon my toy model of evolution with empathy rather than pity.

Let's start with two quotes, from a fine mathematician with unusually broad interests, my late friend Jacob ("Jack") Schwartz, and from a leading theoretical biologist, the inspiring John Maynard Smith. For the full text on the limitations of the mathematical method that I extracted from Jack's provocative essay "The Pernicious Influence of Mathematics on Science," please see the three seemingly contradictory quotes at the begining of this book. Here it suffices to recall that

the mathematical technique can only reach far if it starts
from a point close to the simple essentials of a problem
which has simple essentials.

                                                    —Jacob Schwartz

Coming from someone who has worked in so many areas
of pure and applied math, this is a telling remark.

Perhaps more to the point, here is what Maynard Smith
has to say about mathematical models in biology:

> It may seem natural to think that, to understand a complex
> system, one must construct a model incorporating every-
> thing that one knows about the system. However sensible
> this procedure may seem, in biology it has repeatedly
> turned out to be a sterile exercise. There are two snags
> with it. The first is that one finishes up with a model so
> complicated that one cannot understand it: the point of a
> model is to simplify, not to confuse. The second is that
> if one constructs a sufficiently complex model one can
> make it do anything one likes by fiddling with the param-
> eters: a model that can predict anything predicts nothing.
>
>   —John Maynard Smith and Eörs Szathmáry,
>   *The Origins of Life*

Furthermore, metabiology is in some ways closer to
theoretical physics than it is to pure math. And in theoretical physics highly simplified so-called toy models are
part of the standard tool-kit, part of the standard method-
ology. Hopefully they capture the essential features of a
situation.

Here's another way to put it: We want to find the simplest, mathematically most straightforward situation in which we can prove that life evolves, the minimal system that satisfies the definition of life given in John Maynard Smith's *The Problems of Biology*. Stated differently, we are attempting to find the simplest possible mathematical life-form!

And we most certainly are **not** attempting extremely realistic, highly detailed computer simulations of living systems, which is called "systems biology" and is a popular new field.

I should also make a few remarks about the art of crafting a mathematical theory, remarks that may not be at all obvious to those who have not spent their lives doing mathematics, composing mathematics, immersed in the world of mathematical ideas . . .

Mathematics isn't the art of answering mathematical questions—most questions can't be answered or have ugly, messy, uninteresting answers. Rather math is the art of asking the right questions, the questions that have beautiful, fertile, suggestive answers.

And mathematics isn't a practical tool, a way of getting answers. For that, use a machine, use a computer! Math is an art form, a way to achieve **understanding**! The purpose of a proof is not to establish that something is true, but to tell us **why** it is true, to enable us to understand what is happening, what is going on!

Enough methodological considerations! How does my evolution model actually work? On the following page is a summary of the key ideas that I will now attempt to explain.

Key Ideas of Metabiology
Life = randomly evolving software
A single software organism O, a mathematician!
Biological creativity = Mathematical creativity
Fitness of O = Busy Beaver problem = size of output
Naming big numbers: N, $N + N$, $N \times N$, N^N, $N^{N^{N}}$ N times . . .
Evolution = hill-climbing random walk O, O', O" . . .
Algorithmic mutations $O' = M_1(O)$, $O" = M_2(O')$ . . .
Probability of K-bit algorithmic mutation $M = 2^{-K}$
Non-algorithmic oracles eliminate bad mutations M
Fitness of O increases faster than any computable function!
Thus evolution is creative, not mechanical!

Now we present our toy model of evolution.

Here's the key mathematical idea: Our organisms are mathematicians, and we identify mathematical and biological creativity. Our model is sufficiently abstract that in it there is no essential difference between math and bio creativity.

To keep our organisms evolving, to keep them from stagnating, to avoid a fixed point, we need to challenge them, we need to give them something difficult to do, something that can absorb an unlimited amount of mathematical creativity. To do this we take advantage of Gödel incompleteness (1931), in a form related to Turing's famous halting problem (1936), namely the Busy Beaver problem, which is the problem of naming extremely large integers, extremely large unsigned whole numbers.

The Busy Beaver problem was invented by Tibor Radó, then in his sixties, as we recount in our book Chaitin, da Costa, Doria, *Gödel's Way*. The original reference is: T. Radó, "On Non-Computable Functions," *Bell System Technical J.* **41** (May 1962), pp. 877–884.

Restating the previous point non-technically, we take advantage of Feyerabend's observation in *Against Method* that in science there are no absolutely general methods, which in math is a theorem, the refutation of Hilbert's dream by Gödel (1931) and Turing (1936). As Feyerabend observes, there are no general methods, hence creativity is always required; there is no mechanical way to do science, hence evolution will not stagnate!

My model eliminates physics and eliminates bodies, resulting in a mathematical formulation which is simple enough that we can understand what is going on and prove theorems. Also, I do not worry about the origin of life; I start with fully functioning living beings, and make them evolve forever. I also eliminate populations and eliminate sex: I have only a single organism! And yet it evolves! **You do not need much to make evolution work!** Which is good, because it means that evolution is very basic, very robust!

We are now getting to the mathematical core of our metabiological model of evolution. This is going to be a bit technical, a bit tough. We are going up a steep mountain, but I will try to explain everything in words as best I can, not just with formulas. But if you can't understand something, please just skip it and continue onwards and upwards. Just look at the scenery and try to get an overall feeling for what I am doing.

So we have a single software organism, it's a computer program P, and what interests us is how big a number P calculates. The bigger the number, the fitter P is. We are going to make random changes in P, and there's only one organism P at a time, so P describes what is called a random walk in *software space,* which is the space of all possible software organisms, all possible programs P.

Think of a drunkard staggering about; that's a more colorful name for a random walk, a drunkard's walk. But our organisms P aren't drunkards, they are dedicated mathematicians who are working hard on the Busy Beaver problem, the problem of calculating a really, really large integer.

Technically, we are looking at a special kind of random walk, one that is called a "hill-climbing" random walk. Why? Well, because we are not staggering about completely at random, the fitness has to always increase. We try randomly mutating P until it finally calculates a bigger number. The American population geneticist Sewall Wright gave a colorful name to this kind of a process: he described it in terms of a so-called fitness landscape. If you think of the landscape consisting of the fitness of every possible organism, we are always going uphill, a well-known strategy in many optimization algorithms.

Here's how our hill-climbing random walk works: We try a random mutation, we use it to transform our current organism. If the resulting organism calculates a bigger number, if it is fitter, then it replaces our current organism. Otherwise we stick with the current organism and try another random mutation, and so forth and so on. And the key question is this: How fast does the fitness (= the size of the number that is calculated) grow?

So that's our random walk, and in it there is only one mutating organism. But we are not finished yet. I also have to tell you how you pick the mutations, and with what probabilities.

The crucial step in making metabiology work mathematically is to permit *algorithmic* mutations: If a mutation M is a K-bit program that takes the original organism A as input and produces the mutated organism A' = M(A) as output, then this mutation M has probability 2^{-K}. In other words, if M is a K-bit function, a transformation that can be described in K bits, then it has probability $1/2^K$.

And in our random walk model there is an associated concept of distance, which is measured in bits. An organism A is K bits away from another organism B if the probability that our random walk will carry us from A to B in a single step is $1/2^K$. More formally, the *mutation distance* between organisms A and B is defined to be

$$-\log_2 \text{ of the probability of going from A to B with a single mutation.}$$

It turns out that this is also the size in bits of the smallest program M that takes A as input and produces B as output, which in my previous field, algorithmic information theory, is called *the program-size complexity* of the simplest function M with B = M(A), also known as the *relative information content* of B given A. I mention this because the equivalence between the probabilistic and the program-size way of defining mutation distance is one of the main theorems of my 1975 *J. ACM* paper, which for the first time presented in its full glory the correct formalism for algorithmic information theory.

In other words the mutation distance is the amount of algorithmic information it takes to go from A to B, to get B from A, to *transform* A into B. And please note that it is always possible to get from A to B in a single step. In other words the mutation distance is always finite, the probability of mutating from A to B in a single step is always greater than zero.

Now I have to tell you about the oracle, which is where all the creativity is really coming from in our model. "What oracle?" you may ask. Yes, it's a bit hidden, but it's there in the model I just described.

First of all I should tell you what an oracle is. It's another fun idea from Alan Turing, but not in his famous 1936 paper, in a lesser-known 1939 paper. An oracle is a way for us to compute something that cannot be computed using a normal computer. In particular it can give us a way to decide if a computer program will ever halt. That is called "an oracle for the halting problem."

In other words it's a mathematical fantasy. It's a way to imagine computers that are more powerful than real computers could ever be.

And why do we need an oracle for the halting problem?

We need it to avoid organisms that don't work and mutations that don't work. That is to say organisms that never produce an integer output, and mutations M that when given A never produce a mutated organism B = M(A). There is no algorithm for doing this—it's Turing's unsolvable halting problem—so we need an oracle to tell us when to skip a mutation or an organism that never halts.

So these are the rules of the game, that's how our hill-climbing random walk in software space works. But how fast does it evolve?

We are going to use the so-called *Busy Beaver function*

to measure the rate of biological creativity = the speed of evolution. BB(N) is defined to be the largest integer that can be named in N bits = the biggest integer output produced by any ≤ N bit program that produces a single integer and then halts = the fitness of our fittest ≤ N bit software organism.

In other words, we want to see how fast the fitness grows. And it turns out that it grows very fast, so we need to use something well-known that grows very fast for the purpose of comparison, as calibration. And as Tibor Radó observed, BB(N) grows faster than any computable function of N, faster than N to the Nth to the Nth etc. N factorial times, for example. It will eventually exceed any computable function. From some point on it will always be bigger. We will use BB(N) to characterize the rate of biological creativity in our evolution model.

Why so much emphasis on creativity? Why is creativity so important? Because your parasites, competitors and predators are also evolving! It's an arms race. This is Leigh Van Valen's *Red Queen* hypothesis: You have to run as fast as you can to stay in the same place (which he had to found a new journal to be able to publish)! Genes don't want to be selfish as Dawkins claims in his book *The Selfish Gene*, they want to evolve! That's the reason for sex, which is not at all selfish: With sex you immediately throw away half of your genome! Would you call someone who gives away half of her money selfish? Sex isn't selfish; it improves creativity.

Besides founding a journal so that he could finally publish his most famous paper, Van Valen, who died in

2010, also refused to accept research grants—to preserve his freedom and creativity—and gave higher grades to students who disagreed with him! He really believed in creativity!

Van Valen's *Red Queen* hypothesis is why there is so much creativity in the Earth's biosphere. But in my meta-biological model it emerges by itself from the rules of the game: There are no parasites, competitors or predators; there is only a single organism at a time.

On the exuberant creativity of Nature, see of course Ernst Haeckel's wonderful biology art books. We have been showcasing his work here, at the start of every chapter.

But I do take from Dawkins his emphasis on genes. Who cares about bodies!? This is also the point of Maynard Smith's analysis of life: flames have a metabolism and self-reproduce, but they do not evolve. Life is a system that has heredity and mutations and evolves by natural selection. Flames have no heredity—they do not transmit genetic information, they do not remember how they were started, hence they cannot evolve—which is fortunate, or we would be pursued by giant predatory intelligent fires! Contrast that with my organisms, which have DNA but no bodies and no metabolism.

We are now reaching the climax of the story. For we are now in a position to compare three extremely different evolutionary regimes: brainless exhaustive search, cumulative random evolution and what I call "intelligent design."

You see, to get a feeling for how well Darwinian evolution works in our Busy Beaver model, we need to bracket it from above and below. That way we can see how intelligent, how creative, random evolution actually is.

Here is what the math says:

- **Intelligent design**—which does the best possible, picking successive mutations to try in the best possible way, not at random—reaches fitness BB(N) in time N,
- **Brainless exhaustive search**—which tries all possible organisms at random, that is, does not take the previous organism into account—reaches fitness BB(N) in time 2^N, and
- **Cumulative random evolution**—real Darwinian evolution, which tries all possible algorithmic mutations at random—reaches fitness BB(N) in time between N^2 and N^3.

So you see, cumulative random evolution is much closer to intelligent design than it is to brainless exhaustive search, for N^2 and N^3 are closer to N than to 2^N. This is our *hauptsatz,* our main or fundamental theorem, and it is why we claim that our model evolves and is therefore alive.

As a pure mathematician would now proudly affirm, *QED, quod erat demonstrandum,* which was to be proved. Randomness is creative! Random mutations and natural selection achieve a kind of intelligence! Intelligence emerges spontaneously!

Above I talk about time. But in my model how is that measured? Well,

time = number of mutations that are tried = number of attempted steps in the random walk.

Not all of these mutations succeed, of course, but we count them all.

How does this all work? How do I prove these estimates? Well, in our Busy Beaver toy model of evolution it turns out that the organisms that evolve are better and better approximations to the halting probability Ω—lower bounds, in fact—because these are the fittest organisms. And intelligent design obtains N bits of Ω in time N, which is the best possible; exhaustive search, which stumbles about trying everything, obtains N bits of Ω in time 2^N; and cumulative random evolution obtains N bits of Ω in time between N^2 and N^3.

Remember the previous chapter? There I said that Ω is concentrated mathematical creativity. In other words, we have, respectively, N bits of creativity in time N, and N bits of creativity in time 2^N, and finally N bits of creativity in time between N^2 and N^3.

I won't explain the proof any more here, but I will in the next chapter, which is a talk that I gave at the Santa Fe Institute. Readers with a math or physics background should also take a look at the second appendix, which gives some additional details.

To a non-mathematician these three estimates may not look like much. But I've been trying for forty years— since 1969, look at my long list of publications that mention biology—to find a way to study evolution at this level of mathematical generality, and I could never get to first base. This may not look like much—indeed it is only the first step in the direction of a general, abstract mathematical theory of evolution and biological creativity—but it's that first step that is usually the hardest. What's hard is coming up with concepts that address the fundamental issues and are workable mathematically.

So I am hopeful about this. Yes, at first I was a bit elated, but now the main feeling is cautious hope. Time will tell. I confess that these results do seem a bit strange even to me. After all, it's brand-new, it's a new field. It will take a while to figure out if we are being misled or if there is really something there. Some sort of evolving software model should work, I think.

Now for my Santa Fe Institute lecture, which was, so to speak, the official world debut of metabiology. Let me set the stage. The setting was inspiring. The Santa Fe Institute—a zen-monastery-like structure with magnificent long-range views perched at 2,000 meters above sea level in the New Mexico desert—was surrounded by snow bathed in brilliant sunlight. The audience was ideal: a select group of physicists, mathematicians and computer scientists interested in complex systems. Here is what I told them—by the way this was a few months ago and I had not yet discovered Maynard Smith's 1986 book *The Problems of Biology*.

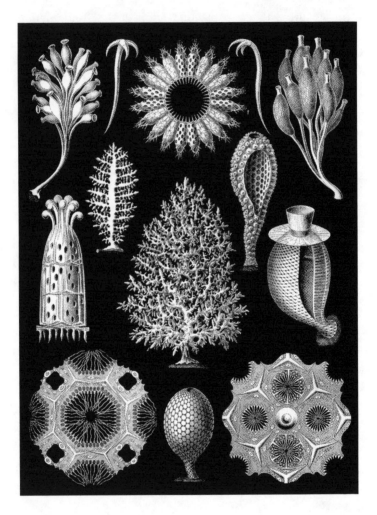

Five

Santa Fe Institute Lecture:
A Mathematical Theory of Evolution
and Biological Creativity

Talk presented Monday, January 10, 2011, at "Random-ness, Structure and Causality: Measures of Complexity from Theory to Applications," a workshop organized by Jim Crutchfield and Jon Machta at the Santa Fe Institute in New Mexico. Material displayed in boldface is what was written on the blackboard.

I want to thank the organizers for inviting me here. I haven't visited the Santa Fe Institute for many years. I'm delighted to be back, and I have something radically new to talk about. I think the time is now ripe to combine theoretical computer science with biology and to begin developing a theoretical mathematical biology.

Theoretical Biology
Mathematical Biology

I believe we already have the mathematical tools to begin developing such a theory.

For many years I have thought that it is a mathematical scandal that we do not have a proof that Darwinian evolution works. I want to find the simplest toy model of evolution for which I can prove that evolution will proceed forever, without ending. My emphasis is on biological creativity, something that has somehow gotten lost in the standard accounts of Darwinian evolution.

I'm aware of the fact that there is a vast literature on biology and evolution—good work, I have nothing against it—but I'm going to ignore most of it and go off in a different direction. Please bear with me.

There is a nice mathematical theory of evolution called "population genetics." But in population genetics by definition there is no creativity, because population genetics **defines** evolution to be changes in gene frequencies in response to selective pressures, and deals with a fixed finite pool of genes. Instead, I am interested in where new genes come from, in how creativity takes place.

Another way to explain my motivation is this: The leading technology of the previous century was based on digital software, computer programming languages. And the leading technology of this new century will be biotechnology, which is based on a natural digital software, namely DNA.

Artificial Digital Software: Programming Languages
Natural Digital Software: DNA

These two technologies will converge. It is no accident that people talk about computer viruses and cyberwarfare and about developing an immune system to protect cyber-assets. And what I am saying is that this isn't just a metaphor. We can take advantage of this analogy to begin developing a mathematical theory of evolution.

Darwin begins his book *On the Origin of Species* by taking advantage of the analogy between **artificial selection** by animal and plant breeders, the successful efforts of his wealthy neighbors to breed champion milk producing cows, racehorses and roses, and **natural selection** due to Malthusian limitations. I want to utilize the analogy between the random evolution of natural software, DNA, and the random evolution of artificial software, computer programs. I call this proposed new field "metabiology," and it studies random walks in software space, hill-climbing random walks of increasing fitness.

Evolution of Mutating Software
Random Walks in Software Space

Random walks are an idea that mathematicians feel comfortable with. There is a substantial literature on random walks. And I am just proposing a random walk in a richer space, the space of all possible programs in a given computer programming language, which is a space that is large enough to model biological creativity.

So I basically start with two observations. First that DNA is presumably what computer scientists call a "universal programming language," which means that it is sufficiently powerful to express **any** algorithm—in particular evo-devo teaches us to think of DNA as a computer program. Second, at the level of abstraction that I am working in my models, there is no essential difference between mathematical creativity and biological creativity, and so I can use mathematical problems for which there are no general methods in order to challenge my organisms and force them to keep evolving.

DNA = universal programming language
Math creativity = biological creativity

Emil Post, who is forgotten but whose work was at the level of that of Kurt Gödel and Alan Turing, considered that the whole point of incompleteness and uncomputability was to show the essential role of creativity in mathematics. The emphasis on formal methods provoked by the computer temporarily obliterated Post's insight, but metabiology picks up the torch of creativity again.

To repeat, the general idea is that we are all random walks in program space! Our genomes are digital software that has been patched and modified for billions of years in order to deal with changes in the environment. In fact, I propose thinking of life as evolving software and considering biology to be a kind of software archeology. Instead of La Mettrie's *L'Homme machine* (1748), we now have *L'Homme software*.

To be more precise, I am studying the following toy model of evolution. I have a single organism, and I try subjecting it to random mutations. If the resulting organism is fitter, then it replaces the original organism.

Now let me explain this in more detail.

What are my organisms? Well, in his book *The Selfish*

Gene, Richard Dawkins teaches us that bodies are unimportant, they are just vehicles for their genes. So I throw away the body and just keep the DNA.

A better way to explain this is to remind you of the definition of life given in John Maynard Smith and Eörs Szathmáry's books *The Major Transitions in Evolution* and *The Origins of Life.* They discuss two definitions of life. The first is fairly obvious. A living being preserves its structure while taking in matter and expelling matter; in other words it has a metabolism. And furthermore it reproduces itself. Although this seems like a natural definition, Maynard Smith and Szathmáry point out that a flame satisfies this definition.

However, flames do not have heredity, fires do not remember if they were started by a match or by a cigarette lighter, and therefore they do not evolve.

Therefore Maynard Smith and Szathmáry propose a more sophisticated definition of life. **You have life when there is heredity with mutations and evolution by natural selection can take place.**

This may seem like a tautology. Darwin's theory may also seem to be a tautology—the survival of the fittest is merely the survival of the survivors—but natural selection is **not** a tautology. And this definition of life isn't either, because the whole point is to prove that there is something that satisfies the definition. The point is to find the simplest system with heredity and mutations that provably evolves.

So to make things as simple as possible, no metabolism, no bodies, only DNA. My organisms will be computer programs. I still have to explain how I do mutations, and what is my fitness measure.

For two years I worked on metabiology using what biologists call point mutations: You change/delete/add one or more contiguous bits in your computer program, and the probability of the mutations decreases exponentially with the number of bits. In this way there is a non-zero probability to go from any organism A to any other organism B in a single mutation, but if all of the bits of A are changed this probability will be infinitesimal.

With point mutations I was able to begin working, I was able to get an idea of what is going on, but the way forward was blocked; things were awkward, the way that pioneering work in math usually is. Then a few months ago, last summer, in July and August, I had a breakthrough.

I realized that from a mathematical point of view the right thing is to consider algorithmic mutations, in which a mutation is a computer program that is given the original organism A and that produces as its output the mutated organism B. If this algorithmic mutation is an N-bit program, then the mutation from A to B has probability 2^{-N}.

N-bit algorithmic mutation A→B has probability 2^{-N}

If we use the prefix-free programming languages of algorithmic information theory, the total probability of all the programs will be less than one, as it should be. This is how to get what is called a "normed measure," and this is a well-known technique.

Now there is again a non-zero probability to go from any organism A to any other organism B in a single mutation step, but the probabilities are very different. Consider, for example, the mutation that flips every bit of the program A. Before, this mutation was possible but extremely unlikely. Now it is a very simple and therefore a highly probable mutation.

If mutations are chosen at random, each mutation will be tried infinitely often, and this bit-flip mutation will be tried very frequently, a fixed proportion of the time in fact.

By the way, with algorithmic mutations it is possible that the mutation program never halts and never outputs the mutated organism B. So you cannot actually simulate our evolution model, because in order to do that you would need to use what computer scientists, following Turing, call an "oracle" for the halting problem. And we will need to use oracles again later on, to decide whether the mutated organism B is fitter than the original organism A. How do we do this? What is our fitness measure, our fitness criterion?

Well, in order to get our organisms to evolve forever, we need to challenge them with a mathematical problem that can never be solved perfectly, that can employ an arbitrary amount of mathematical creativity. Our organisms are mathematicians that are trying to become better and better, to know more and more mathematics. What math problem shall we use to force them to evolve?

The simplest extremely challenging problem is the

Busy Beaver problem, which is intimately related to Turing's famous halting problem. What is the Busy Beaver problem? That's the problem of concisely naming an extremely large positive integer, an extremely large unsigned whole number.

Why does this require creativity? Well, suppose you have a large number N and you want to name a larger number. You can go from N to N + N, to N times N, to N to the Nth power, to N raised to the Nth to the Nth N times. So to name large numbers you have to invent addition, multiplication, exponentiation, hyper-exponentiation, and this requires creativity.

Busy Beaver problem: $N + N$, N^2, N^N, $N^{N^{N^{N^{\cdots}}}}$ (N times)

There is a beautiful essay on the web about this by the quantum computer complexity theorist Scott Aaronson, "Who Can Name the Biggest Number?," which I highly recommend and which explains what a fundamental problem it is.

So that's my fitness measure. Each of my software organisms calculates a single number, a single positive integer, and the bigger the number, the fitter the organism. The current organism A has a particular fitness N, and then we try a random mutation, according to the probability measure that I already explained, and if the resulting organism B calculates a bigger number, then it replaces A. Otherwise we try mutating A again.

Note that again we are implicitly making use of an oracle, because randomly mutating A will often produce a B that never halts and never calculates anything, so that we cannot determine if B is fitter than A—if B produces a number bigger than A does—by merely running B. We need to skip mutations that produce an invalid program B, as well as mutations that never produce a mutated organism B.

And to measure evolutionary progress, to measure the amount of biological creativity that is taking place, the rate of biological creativity, we use the so-called Busy Beaver function BB(N), which is defined to be the biggest positive integer that can be named with a $\leq$ N bit program. [This is a more refined version of the Busy Beaver function. The original Busy Beaver function BB(N) was the biggest integer calculated by a Turing machine with $\leq$ N states.]

BB(N) = largest positive integer named in $\leq$ N bits

BB(N) grows faster than any computable function of N, because BB(N) is essentially the same as the longest run-time of any $\leq$ N bit program that eventually halts. So if BB(N) were computable that would give us a way to solve the halting problem.

Okay, now let's see what happens if we start with a trivial organism, for example the one that calculates the number 1, and we carry out this hill-climbing random walk. We apply mutations at random and look how fast the fitness will grow. In fact, to calibrate how fast cumulative random evolution will work, let's see where it falls between

- brainless **exhaustive search**, in which the previous organism A is ignored and we try a new organism B at random (in other words, the mutations are produced by programs that are chosen at random as before, but that are not given any input),
- and the fastest possible evolution that we can get by picking a computable sequence of mutations in the best possible manner in order to make the fitness grow quickly, which I call "**intelligent design.**"

    [You cannot use an oracle to jump directly to BB(N), BB(N+1), etc., because we are only allowing a highly restricted use of oracles, in determining whether A→B is fitter than A. Furthermore, to eliminate mutations that don't produce a B from A.]

The designer isn't the deity, he is the mathematician who finds the best possible sequence of mutations to try.

Here is what happens with these three different evolution regimes:

Exhaustive Search **reaches fitness BB(N) in time** $\approx 2^N$
Intelligent Design **reaches fitness BB(N) in time N**
Cumulative Evolution **reaches fitness BB(N) in time between N^2 and N^3**

My unit of time is the number of mutations that are tried. For example, I try one mutation per second. Note that the fitness increases faster than any computable function of the time, which shows that genuine creativity is taking place. If the organisms were improved mechanically, algorithmically, then the fitness would only grow as a computable function of the time. Of course the creativity is actually coming from the implicit use of an oracle: Each time we try a mutation and are told if the resulting B is fitter than the original organism A, we get at most one more bit of creativity and we can advance from BB(N) to BB(N + 1). That is the best we can do, and that is what intelligent design accomplishes.

Exhaustive random search takes time that grows exponentially in N to get to BB(N), because exhaustive search is ergodic, it is searching the entire space of possible organisms. That is not at all what happens in real evolution: The human genome has 3×10^9 bases, but in 4 billion years the biosphere has only been able to try an infinitesimal fraction of the astronomical number $4^{3 \times 10^9}$ of all possible DNA sequences of that size. Evolution is not at all ergodic.

Note that in our toy model cumulative evolution is much faster than exhaustive search, and fairly close to intelligent design. How come? In fact, what is happening in this random evolution model is that we quickly evolve very good lower bounds on the halting probability Ω. Knowing the halting probability Ω with N bits of precision is essentially the same as knowing BB(N). And the random mutations M_K that rapidly increase the fitness are ones that take a lower bound on Ω and see if they can add a 1 in the Kth bit position after the decimal point. In other words, M_K tries to see if the current lower bound on Ω is still a lower bound if we add 2^{-K}.

M_K: Can we add 2^{-K} to our lower bound on Ω?

If so, we add 2^{-K} to our lower bound on Ω. If not, we try incrementing another bit K at random. Intelligent design systematically tries M_K for K = 1, 2, 3, . . . Cumulative evolution is not much slower, because mutation M_K essentially only needs to name K, which can be done with log K + 2 loglog K bits of prefix-free/self-delimiting software, and therefore has probability $\geq 1/K(\log K)^2$ and will happen in time expected to be $\leq K(\log K)^2$. So cumulative evolution will try incrementing bits K = 1 through N of Ω in order by time roughly

$$\sum_{K \leq N} K(\log K)^2 \ \leq \ \text{between } N^2 \text{ and } N^3$$

This is an outline of the proof that Darwinian evolution works in my model. For the details please see the second appendix or my University of Auckland Centre for Discrete Mathematics and Theoretical Computer Science Research Report CDMTCS-391 at this URL:

http://www.cs.auckland.ac.nz/CDMTCS// researchreports/391greg.pdf

I admit that this result seems a bit strange even to me, but I think that it **is** a first step in a metabiological direction. It is the simplest model that I can think of where you can prove that evolution works. It's my attempt to extract the mathematical essence of Darwin's theory. To my surprise, the organisms that rapidly evolve in this model are better and better lower bounds on the halting probability Ω. In fact, the halting probabilities of all possible universal Turing machines are rapidly evolving in parallel; there are actually many halting probabilities, not just one. We will know N bits of each Ω in time roughly between N squared and N cubed.

Why is Ω the organism that evolves? Well, it's because a key thing in Darwin's theory is that evolution results from accumulating small changes each of which is advantageous. Darwin worried that half an eye was useless and was very concerned with the absence of intermediate forms. A chapter in his book *On the Origin of Species* is "On the Imperfection of the Geological Record." Ω is jam-packed with useful mathematical information and we can learn one bit of its numerical value at a time, so that better and better lower bounds on Ω give us a highly probable evolutionary pathway by summing small, nevertheless advantageous, changes.

As I said before every mutation is tried infinitely often, and some are pretty violent. There is a mutation that replaces an organism with fitness N by a program that directly outputs N + 1 without doing any computation. This is a pretty stupid organism, but it increases the fitness, and so this mutation is successful whenever it is tried. How come random evolution works in spite of such violent perturbations? Well, that's because the memory of the system resides in the fitness, which always increases. Knowing a very large integer N enables us to calculate a very good lower bound on Ω. Just look at all the programs up to N bits in size and run them for time N to see which ones halt, and that gives you a better and better lower bound on Ω.

So this is my current best effort to find the Platonic ideal of evolution, the simplest, most natural system that exhibits creativity and that I can prove evolves by random natural selection. We get provable evolution, which is a good first step, and which I think validates metabiology as a possible research program, but we fail to get an increase in hierarchical structure in our organisms—which are essentially just lower bounds on Ω—and hierarchical structure is a very conspicuous feature of naturally-occurring organisms.

What about hierarchical structure?

I actually have two more toy models of evolution that I have studied, not just the one I have explained. What varies in these models is the fitness measure, and also the programming language. In my second model I use what is called a "subrecursive" programming language, one that is not universal and for which there is no halting problem. There is no halting problem because this is a FORTRAN-like language in which each time you enter a loop you know in advance exactly how many times it is going to be executed.

And now each program calculates a function, not an integer, and the faster the function grows, the fitter the program.

$$N + N \rightarrow N^2 \rightarrow N^N \rightarrow N^{N^{N^{N^{\cdots}}}} \quad \textbf{(N times)}$$

A lot is known about subrecursive hierarchies (see for example the book by my friend Cristian Calude, *Theories of Computational Complexity*), and using all of this it is easy to show that the loop-nesting level of the programs must increase without bound. So I also have a toy model of evolution in which hierarchical structure provably emerges.

In my third toy model of evolution, the programs are once more universal, not subrecursive, and each program names what is called a "constructive Cantor ordinal number." Here are some examples of such numbers:

$$1, 2, 3, \omega, \omega + 1, \omega + 2, 2\omega, 3\omega, \omega^2, \omega^3, \omega^\omega, \omega^{\omega^\omega}, \varepsilon_0 \cdots$$

In this model I conjecture that exhaustive search is the best that you can do. In general, you expect that with arbitrary fitness landscapes an exhaustive search will be needed and you will not get cumulative evolution. The fitness landscape has to be very special for Darwinian evolution to work.

So where does metabiology go from here? I expect that there is a spectrum of possible models of randomly evolving programs. More realistic models will limit the run time of programs and thus avoid the need for oracles. I expect there to be a trade-off between biological realism and what can be proved: The more realistic the model, the more we will have to rely on computer simulations rather than proofs.

Are there more realistic models?

There are many possibilities for future work. Besides limiting the run time, one can try to incorporate populations or sex. Much remains to be done. But one shouldn't expect this theoretical mathematical biology to ever become as realistic as theoretical physics. Biology is just too messy, too far removed from mathematics. And although metabiology is promising mathematically, it remains to be seen how relevant metabiology will ever be to real biology. But as my wife, Virginia Chaitin, points out, metabiology has already raised an interesting question for real biologists, which is how powerful are mutational mechanisms in real organisms? How closely do real organisms approach the powerful algorithmic mutations needed to make my metabiological models work?

(In fact, the writings of Maurício Abdalla and Máximo Sandín have subsequently suggested to me that algorithmic mutations may actually exist: they are viruses!)

Another caveat about metabiology is that it does not study the origin of life nor does it say anything about what may happen if we begin to take charge of our biological destiny by doing genetic engineering and producing children with designer genomes—with the best genes that money can buy.

I'd like to end with a few general remarks about biological creativity and evolution.

The conventional view is that evolution is not unceasing; you adapt perfectly to your environment, and then you stagnate. And people claim that evolution is not about progress. The simple mathematical models that people have built up to now, in biology and in economics, talk about stability and fixed points, they do not talk about creativity.

But that is not the right way to think about biology. In biology nothing is static, everything is dynamic. Viruses, bacteria and parasites are constantly mutating, constantly probing, constantly trying to find a better way through your defenses, constantly running through all the combinatorial possibilities. Biology is ceaseless creativity, not stability, not at all. It's an arms race, and as Lewis Carroll's Red Queen said, you have to run as fast as you can to stay in the same place.

(For a magnificent visual depiction of the exuberant creativity of Nature, see Ernst Haeckel's splendid books *Art Forms from the Ocean* and *Art Forms in Nature*.)

This point is particularly well illustrated by the so-called paradox of sex that is discussed at length in the section on the rotifer in Dawkins' *The Ancestor's Tale*. In the standard view of Darwinian evolution, sex is problematic because supposedly selfish genes just want to copy themselves. But with sex you immediately throw away half of your genome, which is not at all selfish—would you call a person who gives away half of her money selfish?! Nevertheless there are very few parthenogenetic species and sex is almost universal. How come?

Why is there sex?

The reason is that biology is actually all about constant creativity and change; nothing is stable, just like in human affairs. And sex greatly speeds up creativity. If there are several needed mutations, sex takes the maximum of the time needed for each to occur randomly in order to get them all, whereas parthenogenetically it takes the sum of the expected mutation times instead of the maximum.

In summary, metabiology emphasizes biological creativity, not selfishness, and it opens the door to a completely new interpretation of Darwinian evolution. It remains to be seen how far this path will lead, but the first steps are encouraging. The mathematical tools are now in place to study the evolution of mutating software. Theoretical computer science **is** theoretical biology.

FURTHER READING

S. Aaronson, http://www.scottaaronson.com/writings/bignumbers.html.

C. Calude, *Theories of Computational Complexity,* Elsevier, 1987.

G. Chaitin, *Algorithmic Information Theory,* Cambridge University Press, 1987.

G. Chaitin, "Evolution of Mutating Software," *EATCS Bulletin* **97** (February 2009), pp. 157–164.

G. Chaitin, "Metaphysics, Metamathematics and Metabiology," *APA Newsletter on Philosophy and Computers* **10**, No. 1 (Fall 2010), pp. 7–11. Also in H. Zenil, *Randomness Through Computation,* World Scientific, 2011, pp. 93–103.

G. Chaitin, http://www.cs.auckland.ac.nz/CDMTCS//researchreports/ 391greg.pdf. Also in H. Zenil, *A Computable Universe,* World Scientific (forthcoming).

C. Darwin, *On the Origin of Species,* John Murray, 1859.

R. Dawkins, *The Selfish Gene,* Oxford University Press, 1976.

R. Dawkins, *The Ancestor's Tale,* Houghton Mifflin, 2004.

E. Haeckel, *Art Forms from the Ocean,* Prestel, 2009.

E. Haeckel, *Art Forms in Nature,* Prestel, 2010.

J. O. de La Mettrie, *Man a Machine,* Open Court, 1912.

J. Maynard Smith and E. Szathmáry, *The Major Transitions in Evolution,* Oxford University Press, 1997.

J. Maynard Smith and E. Szathmáry, *The Origins of Life,* Oxford University Press, 1999.

Six

Theological Aspects of Metabiology

Okay, in the previous chapters we've looked at the math, but what does it all mean? Assuming it's right, what does it tell us about *la condition humaine*? In the final three chapters we discuss the theological, political and episte-mological implications of metabiology. Let's start with theology . . .

First of all, I want to emphasize that metabiology is not an atheistic theory. In my opinion, the attempt to find a beautiful mathematical theory for biology springs from a religious impulse: Leibniz's theology, that the laws of our universe must be beautiful, because—in my words, not Leibniz's—the world is a work of art created by a God who is a mathematician. Remember, Leibniz created a version of theology that was compatible with modern science and mathematics. I view metabiology as religious in that sense, in the sense of Leibniz and Spinoza. I'm trying to get a plastic world out of conventional science and mathematics. Like Leibniz, I want to have my cake and eat it too.

Can Man comprehend God? The gnostics—from the Greek word for knowledge—thought so, the agnostics thought God was unknowable. Leibniz believed God could be comprehended through all-powerful Reason, while others thought He could only be reached through blind Faith, he could never be fully comprehended. Here is a modern version of this: Can Man ever fully comprehend the laws of the Universe? A relic of Leibniz's belief in all-powerful Reason is the contemporary doctrine that when something is fully understood it can be expressed mathematically, that the more mathematical a science is, the more it has progressed.

Let's recall what Leibniz has to say about simplicity and complexity. According to Leibniz, *Discours de métaphysique,* Sections V and VI (1686), the world is less perfect if God has to intervene to create Life. A world requiring miracles is less perfect than one that runs by itself, because it is more complicated, because all is not derived from the simple basic laws of nature. Miracles are like amendments that have to be added to the fundamental laws of the land, greatly increasing the size of the constitution. This is basically Leibniz and Malebranche's argument against miracles. (See Steven Nadler, *The Best of All Possible Worlds: A Story of Philosophers, God, and Evil,* pp. 132–133. See also Matthew Stewart, *The Courtier and the Heretic: Leibniz, Spinoza, and the Fate of God in the Modern World.*)

As a working mathematician, I'm afraid I do not have a fully worked out coherent philosophical position that I am prepared to defend. I am a Pythagorean in the sense that I believe that the world is built out of mathematics, that the ultimate ontological basis of the universe is mathematical, which is the hardest, sharpest, most definite possible substance there is, static, eternal, perfect. (After all, the world must be made out of something, and we certainly can't use marshmallows!) This is more or less the position of modern theoretical physics, it seems to me.

I guess that I do not share the Platonist view that this world is but an imperfect version of the world of ideas. Rather, to my way of thinking, our world is built atop that Platonic world. I don't think I am a dualist, I don't think I believe in a separate world of mathematical ideas, a world apart from our own. Rather it seems to me that our physical world is but an infinitesimal portion of the world of mathematical ideas, which includes all possible physical universes, and which is all that exists, all that really is . . . But, following Gödel, our knowledge of that perfect world is always incomplete, always partial, and constantly changing.

In my picture of it, the Platonic world of mathematical ideas simultaneously contains all possible physical universes and all possible ways of doing mathematics. At the beginning of the final chapter, which is on the ultimate limits of mathematics, I say that math is constantly evolving; that is our mathematics, human mathematics, math with a small m, but Mathematics with a capital M, following Plato, I think of as static, eternal and perfect.

Please compare Max Tegmark's views. (See M. Tegmark, "Parallel Universes," in J. D. Barrow, P. C. W. Davies and C. L. Harper, *Science and Ultimate Reality: Quantum Theory, Cosmology and Complexity,* Cambridge University Press, 2004, pp. 459–491.) Tegmark posits an extreme multiverse consisting of all possible mathematical laws each "creating" or associated with a universe governed by those laws.

The last remnant of belief in God is the belief in beautiful scientific theories. A century ago educated people worshipped Beauty, Truth and Art—and then they were always capitalized. Nowadays we believe in nothing. Or maybe just in money. The Enlightenment has ended in a *reductio ad absurdum*.

Metabiology is not an atheistic theory. By making biology Pythagorean, metabiology sides with Reason, with Truth and with Beauty. And with Leibniz's God and Spinoza's God and Einstein's God.

Furthermore the picture of Nature presented by metabiology is not mechanical and reductionist, it is not a closed Hilbertian formal system. Instead the Natural World, by creating software, becomes plastic, creative and open; it evolves forever; it never stagnates. This is in sympathy with Henri Bergson's *L'Evolution créatrice* (1907), with George Bernard Shaw's "life force"—see the "Don Juan in Hell" act of his play *Man and Superman*—and with Gilles Deleuze's *devenir.*

A Note on Randomness: Darwin replaces God by randomness. This may seem bad, but in fact randomness is creative; it is liberating; it is better than being in a mechanical, deterministic universe, which is just a giant prison. A capricious God is infinitely better than a boringly predictable God. Creativity is by definition surprising, capricious, unpredictable.

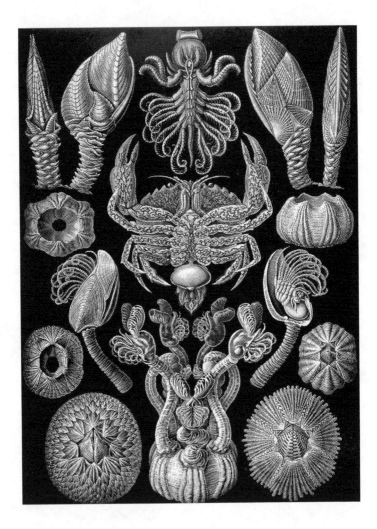

Seven

The Politics of Creativity
(Social Darwinism → Social Metabiology)

According to metabiology the purpose of life is creativity, it is not preserving one's genes. Nothing survives, everything is in flux, *ta panta rhei,* everything flows, all is change, as in Heraclitus.

Please consider these examples of extreme and mysterious creativity: Bach, Mozart, Euler, Cantor, Ramanujan.

Leonhard Euler wrote a beautiful paper every week during his long life, always on a different subject, even when he was blind. He makes creativity look effortless. He explains his entire train of thought; while reading him you think you could do it just as well, but the illusion lasts only as long as you are reading him. He seems to have had some kind of direct access to the source of creativity; for him it was a flood, no publisher could keep up . . .

Georg Cantor created a kind of mathematical theology, a mathematics of the infinite, the transcendent, as a way to attempt to reach God in the limit. His paradoxical theory reflects the paradox of a finite being attempting to comprehend the infinite, of Man trying to comprehend a transcendent God. Nevertheless it is an inspiring theory that had an immense effect on twentieth-century mathematics, promoting a much more general set-theoretic structural view . . .

Srinivasa Ramanujan achieved his most beautiful results without proof, by some kind of inexplicable insight. He claimed that the goddess Namakkal—also called Namagiri—of Tamil Nadu would write equations on his tongue as he slept. Furthermore, he proclaimed that no equation is of value unless it expresses one of God's thoughts . . .

Take that, you reductionists!

I propose that we should measure a nation by its creativity, by its production of new scientific, artistic, technological or social ideas, not in terms of money. And what can a nation do to maximize its creativity?

Please note that Feyerabend's *Against Method* is actually a book on politics. Because societies impose method, they impose rigidity. A scientist cannot be flexible while imbedded in a rigid society . . .

Nations die when they become rigid, when their bureaucracy overwhelms everything, when they stagnate, when they become inflexible, when they think too much of themselves. At a certain point, prosperity can become a problem. With great effort one can overcome poverty and make something of oneself, but it is almost impossible to overcome excessive affluence.

Furthermore, in my opinion, human beings should not try to be like machines; machines are much better at that than we are. Human beings excel at being creative, at being as un-machine-like as possible.

Look at how creative ancient Greece was, and how boringly stable ancient Egypt was. For the Greeks were separated by islands and mountains into city-states small enough that unique individuals could have an impact, while Egyptian geography made possible strong central control over a large territory that suppressed all change.

Or look at Renaissance Italy, so creative and so divided into little duchies and principalities. You will not be surprised to hear that I advocate creative anarchy and decentralized control.

Remember, computer technology emerges from Turing's desire to determine the scope of Hilbert's formal axiomatic theories, and universal manufacturing may emerge from von Neumann's attempt to understand self-reproduction. Would either of these research projects be funded now? Under the current funding regime, is it possible to obtain support for long-term basic research with little apparent prospect of practical applications?

Michael Faraday (1791–1867), when asked by a politician what good were his electrical discoveries, what good was electricity, replied, "Of what use is a baby? Someday you can tax it!"

Carl Jacobi (1804–1851), when asked why he worked on pure mathematics, replied, "For the honor of the human spirit!"

Now only safe, incremental "normal science" (Thomas Kuhn) is funded and published, not paradigm shifts. Creativity is slaughtered.

A good example of this is provided by the career of Leigh Van Valen, discoverer of the fundamental *Red Queen* hypothesis, which explains biological creativity and sex, and which was rejected for publication many times. To maintain his freedom of action, Van Valen refused to apply for research grants. Would anyone dare to behave this way now?

And how did the author of *Proving Darwin* manage, you may ask. Well, for many years I did practical computer hardware and software engineering, and worked on my theoretical interests as a hobby.

Another example of the way the current system works is provided by considering the fate of Julian Schwinger's paper, or papers, on mechanisms that could possibly account for Pons and Fleischmann's cold fusion experiments. In spite of having a Nobel Prize, Schwinger was unable to get his paper accepted for publication in any reputable physics journal, and angrily remarked, "These people are forgetting that physics is also an experimental science!," or words to that effect.

More recently the Italian inventor Andrea Rossi has had, like Van Valen, to create his own journal in order to publish his work on a practical device for what are now called low-energy nuclear reactions. This device, which he demonstrated to the press at the University of Bologna in January 2011, apparently fuses hydrogen and nickel obtaining copper and heat, and unlike Pons and Fleischmann's original device, seems to be inexpensive and entirely practical, reliable and safe. Rossi says he intends to ship commercial units later in 2011. If this actually comes to pass, take that, factory science; take that, the referee and grant systems! For as Richard Feynman remarked, Nature's imagination is far greater than ours.

The creative spirit will survive, factory science will not

triumph, for one simple reason. Societies that suppress creativity temporarily gain increased stability, increased efficiency, but they are not flexible enough to deal with a changing environment. Ultimately they fall by the wayside, like the dinosaurs, that were replaced by the smarter, more flexible mammals. Human beings are not especially good at anything in particular, but we are very curious and extremely adaptable. Like the universal Turing machine, we are generalists, not specialists.

To survive, a society needs to impose some coherence, but not too much, lest it do away with creativity altogether. It is a delicate balance, of permitting some individuals to break the rules, up to a point. We are clearly going in the wrong direction now in some societies where creativity is micromanaged by gigantic bureaucracies.

In my view, our most urgent task today is to be creative enough to design a flexible society, a society in which creativity is somehow tolerated, not like Aldous Huxley's *Brave New World,* which eliminated art and intelligence in favor of stability:

We must be creative enough to design a society that permits creativity!

Some remarks on this stemming from my Jewish cultural heritage. The Talmud contains endless discussion and argument, and a rabbi's authority as the leader of a shtetl was based only on his knowledge and intelligence, it was not hereditary. Every male is supposed to study; everyone has an opinion. As in Hindu culture, there is no central authority; gurus have their own way of teaching, even their own versions of the fundamental truths . . .

In line with this tradition, Israel is, it seems to me, a rather creative, argumentative society; it is forced to be creative by the constant state of war. Army buddies, perhaps from intelligence, often leave the army and form high-tech startups. People move from startup to startup . . .

Let me end this chapter with some quotations from G. H. Hardy, *A Mathematician's Apology* (Cambridge University Press, 1940), which inspired many math students like myself:

> It is never worth a first class man's time to express a majority opinion. By definition, there are plenty of others to do that.

> A mathematician, like a painter or a poet, is a maker of patterns. If his patterns are more permanent than theirs, it is because they are made with ideas.

> Nothing I have ever done is of the slightest practical use.

Eight

What Can Mathematics Ultimately Accomplish? Metabiology and Beyond

What a surprise that we have been able to develop a first cut at a fundamental mathematical theory for biology. Just a few years ago I would have thought this to be impossible, a mad dream, a three-hundred-year project. Instead, given a little bit of inspiration, just a few good ideas, it turned out to be a three-year project. First, what are possible future directions for metabiology? But most of all, what else that now seems completely beyond reach might math ultimately be able to achieve?

Math itself evolves, math is completely organic. I am not talking about what Newtonian math might ultimately be able to achieve, nor what modern Hilbertian formal axiomatics might ultimately be able to achieve (see Jeremy Gray, *Plato's Ghost: The Modernist Transformation of Mathematics,* Princeton University Press, 2008), and not even what our current postmodern math might ultimately be able to achieve. Each time it faces a significant new challenge, mathematics transforms itself. Kuhnian paradigm shifts are not limited to the experimental sciences, they also take place in mathematics, supposedly an *a priori* discipline, a necessary tool of thought.

And as Max Planck said, science advances one death

at a time. More precisely, he said that new scientific ideas
are never accepted by their opponents. Instead what happens is that a new generation grows up with the new
ideas, and subsequently takes them for granted. Here is
the actual wording in Planck's scientific autobiography:
"A new scientific truth does not triumph by convincing
its opponents and making them see the light, but rather
because its opponents eventually die, and a new generation grows up that is familiar with it." (Cited in Thomas
Kuhn, *The Structure of Scientific Revolutions.*)

So mathematics is constantly changing. What is considered a valid proof is constantly changing. There is
even an empirical mathematics based on computational
evidence instead of proofs (Jonathan Borwein and Keith
Devlin, *The Computer as Crucible: An Introduction to
Experimental Mathematics,* A K Peters, 2008), one that
following Richard Feynman might be described as Baby-
lonian rather than Greek in style (Feynman, *The Character of Physical Law,* MIT Press, 1965).

So the mathematics of the future may be unrecogniz-
able. What about metabiology? Let's discuss the future of
metabiology and possible metabiological research topics
and projects.

We have a definition of life (Maynard Smith, 1986)
plus a mathematical proof that something exists satisfying the definition (2010). This is elegant but seems far
removed from conventional biology. Can we do better?

One possibility is to do computer experiments instead
of proving theorems: *experimental metabiology,* computer experiments run on clusters of machines. For this

experimental approach to work we must limit program run times so that no oracles are needed.

We can also try using non-universal programming languages for our software organisms, limited programming languages which don't have a halting problem.

Or perhaps we can try using computer programming languages with a more biological flavor, ones that support parallel pattern-matching for instance. And how about studying biology-inspired mutations, like duplicating a subroutine, something which happens with genes. Once you have copied an important gene, mutations in one of the copies are okay, nothing essential is lost.

How about complicating our random walk model with populations or sex? What is the best way to fit these into our current formalism?

Let's get back to *theoretical metabiology,* that is, to proving theorems. The current version of metabiology is based on *computability theory* and the distinction between the computable and the uncomputable. The oracle provides the uncomputable "divine inspiration" that enables our mathematician organisms to evolve, to improve themselves, to become substantially smarter. I expect that one could also use *time complexity theory.* Instead of distinguishing between what is computable and what is uncomputable, time complexity theory distinguishes between what can be computed quickly, and what requires a substantially greater amount of time.

I expect that there will be a trade-off: *The more realistic a version of metabiology is, the less one will be able to prove.*

Returning therefore to more practical metabiological models, ones that I expect will have to be explored experimentally rather than theoretically, here are some ideas about how to make collecting empirical data about randomly evolving software more practical:

• First of all, to speed up evolutionary experiments, it is important to avoid considering mutations that produce programs that have obvious errors or are clearly equivalent.

• Second, we can take advantage of software engineering practice. In large software projects, techniques such as object-oriented programming with encapsulation and levels of abstraction are commonly used to produce maintainable code and keep it that way in spite of constant updates. Such techniques prevent "spaghetti" code and keep the changes required to fix bugs or enhance function localized. It is much better if a change can be made in one spot instead of requiring simultaneous updates scattered throughout the code. This is also helpful if code is evolving by random mutation instead of by human intervention. It increases the probability that a random mutation will be useful.

Techniques like this should enable computer experiments to be more productive by increasing the rate of evolutionary progress. With a little bit of luck, metabiology will simultaneously develop theoretically and experimentally.

Now let's talk about something else.

If people do not believe that mathematics is simple, it is only because they do not realize how complicated life is.

—John von Neumann

This book is dedicated to von Neumann, whose Hixon Symposium paper in a sense founds metabiology, and who inspired me as a student. I read everything written by him that I could get my hands on, including two brilliant essays excerpted in James R. Newman's fabulous multi-volume anthology *The World of Mathematics* (1956). From von Neumann I learned that Turing's 1936 paper creates computer hardware and software, at least as a mathematical idea.

I learned from von Neumann that mathematics could be applied everywhere, that math applies to everything: theory of games, automata theory, the Hilbert space formalism for quantum mechanics... Von Neumann would create a new field of mathematics every day before breakfast, or so it seemed to me. He made it look easy.

And for a number of years von Neumann and Morgenstern's massive, brilliant *Theory of Games and Economic Behavior* (1944) was a frequent companion. In this amazing book von Neumann presents his entire train of thought, he shows us how to create a new field.

The theory of games can even be viewed as a mathematical theory of ethics and morality, or at least it leads in that direction. This was clear to me many years ago, even though I never worked on it, and it is a point of view developed at length in Martin Nowak's recent book

Supercooperators. So why not a mathematical theory of *beauty,* of *thinking,* of *consciousness,* of *psychology,* of *anthropology* (possible social arrangements, possible social structures), of *historical dynamics* (recall Isaac Asimov's *Foundation Trilogy,* 1951–1953)? I immediately got to work, as a teenager, to create a mathematical theory of randomness as lack of structure, as incompressibility.

The proximate cause, the spark, was reading a footnote in von Neumann and Morgenstern on the fact that quantum mechanical randomness was necessary in order to be able to formulate a theory for zero-sum games without a saddle point. It seemed strange to me that such a theory would not be possible in a classical, deterministic world. So I remembered and got to work developing an idea I had previously had about defining binary sequences lacking any structure. Such sequences, I reasoned, would surely work for playing zero-sum games such as matching pennies even if they were not produced by a quantum mechanical system . . .

A recent essay in arxiv.org suggests that sets that can be members of themselves, reflexive sets x with x ε x, non-well-founded sets, have something to do with the self-referential nature of consciousness. I suspect this idea is not deep enough, not revolutionary enough, but it is something. See Willard Miranker and Gregg Zuckerman, "Mathematical Foundations of Consciousness" (http://arxiv.org/pdf/0810.4339), and the lovely book review by Martin Gardner, "Do Loops Explain Consciousness?" (*AMS Notices* **54** (2007), pp. 852–854).

Furthermore, a very simple game, the iterated Prisoner's Dilemma, has been used to great effect to study the evolution of cooperation. Many people have worked on this: Anatol Rapoport, Robert Axelrod, Karl Sigmund, Martin Nowak . . . This game has the following payoff matrix:

(Payoff to A, to B)	Prisoner B cooperates	Prisoner B defects
Prisoner A cooperates	$(-2, -2)$	$(-4, -1)$
Prisoner A defects	$(-1, -4)$	$(-3, -3)$

The unit of the payoffs is years in prison (a negative payoff). A and B play this game again and again. One simple strategy is "Tit for Tat." Do whatever your opponent did in the last play. If your opponent cooperated, so do you, otherwise don't. Another strategy, discussed at length by Nowak, is "Win Stay, Lose Shift." If you did well on the previous play, repeat your previous choice, otherwise change your previous choice. More precisely . . . but maybe you should get the details from Nowak and Highfield, *Supercooperators*.

The universe is not only stranger than we imagine, it is stranger than we can imagine.

—Haldane

(As often happens, this is a simplified version of the original: "I have no doubt that in reality the future will be vastly more surprising than anything I can imagine. Now my own suspicion is that the Universe is not only queerer than we suppose, but queerer than we *can* suppose." J. B. S. Haldane, *Possible Worlds and Other Papers,* Chatto & Windus, 1927, p. 286.)

What about the ultimate future of science and mathematics? Allow me to offer you my so-called *extended Copernican principle*: The Earth isn't the center of the universe, we are not at a unique position in *space,* and therefore we are also not at a unique position in *time.* I see no reason why our current science should be anywhere near final, why our current scientific ontology will survive. Perhaps, to use Wolfgang Pauli's telling putdown, it is not even wrong.

Rather than supposing that our current science is essentially final, I prefer to use linear extrapolation: A century ago there were no electrons, fifty years ago there were no computers. Who knows what we will know fifty or a hundred years from now? Four hundred years ago Newton had not yet been born, and four hundred years from now modern science will be twice as old and wise as it is today. Perhaps then it will contain "electron psychology," an intriguing phrase in a lovely science fiction story by A. E. van Vogt that I read as a child.

I think that science and magic are not that different. They both believe in a hidden reality behind everyday appearances. Anyway, as Arthur C. Clarke said, any sufficiently advanced technology is indistinguishable from magic.

Furthermore, in my opinion quantum mechanics and quantum information theory are not materialist, they are already a complete paradigm shift: the world as idea, as information, not matter. We don't have to wait long for a totally new ontology; it has already happened.

Let me close with the wonderful opening quote in Karl Popper's *The Logic of Scientific Discovery*:

Theories are nets: only he who casts will catch.

—Novalis

The details of the particular "evolution of mutating software" models presented in this book are not that important. I've heard a biologist say that all the details in Schrödinger's book *What Is Life? The Physical Aspect of the Living Cell* are wrong. Nevertheless, *What Is Life?* raised the issue, it suggested to many people that the time was ripe to create molecular biology. I could not hope for this book to have a better fate. Even if almost everything in this book is wrong, I still hope that *Proving Darwin* will stimulate work on mathematical theories of evolution and biological creativity. The time is ripe for creating such a theory.

Von Neumann's "DNA = Software" Paper*

THE CONCEPT OF COMPLICATION; SELF-REPRODUCTION

The Concept of Complication

The discussions so far have shown that high complexity plays an important role in any theoretical effort relating to automata, and that this concept, in spite of its *prima facie* quantitative character, may in fact stand for something qualitative, for a matter of principle. For the remainder of my discussion I will consider a remoter implication of this concept, one which makes one of the qualitative aspects of its nature even more explicit.

There is a very obvious trait, of the "vicious circle" type, in nature, the simplest expression of which is the fact that very complicated organisms can reproduce themselves.

We are all inclined to suspect in a vague way the existence of a concept of "complication." This concept and its putative properties have never been clearly formulated.

---

*From John von Neumann, "The General and Logical Theory of Automata," in Lloyd Jeffress, *Cerebral Mechanisms in Behavior: The Hixon Symposium,* John Wiley and Sons, 1951, pp. 1–41.

We are, however, always tempted to assume that they will work in this way. When an automaton performs certain operations, they must be expected to be of a lower degree of complication than the automaton itself. In particular, if an automaton has the ability to construct another one, there must be a decrease in complication as we go from the parent to the construct. That is, if A can produce B, then A in some way must have contained a complete description of B. In order to make it effective, there must be, furthermore, various arrangements in A that see to it that this description is interpreted and that the constructive operations that it calls for are carried out. In this sense, it would therefore seem that a certain degenerating tendency must be expected, some decrease in complexity as one automaton makes another automaton.

Although this has some indefinite plausibility to it, it is in clear contradiction with the most obvious things that go on in nature. Organisms reproduce themselves, that is, they produce new organisms with no decrease in complexity. In addition, there are long periods of evolution during which the complexity is even increasing. Organisms are indirectly derived from others which had lower complexity.

Thus there exists an apparent conflict of plausibility and evidence, if nothing worse. In view of this, it seems worthwhile to try to see whether there is anything involved here which can be formulated rigorously.

So far I have been rather vague and confusing, and not unintentionally at that. It seems to me that it is otherwise impossible to give a fair impression of the situation that exists here. Let me now try to become specific.

Turing's Theory of Computing Automata

The English logician, Turing, about twelve years ago attacked the following problem.

He wanted to give a general definition of what is meant by a computing automaton. The formal definition came out as follows:

An automaton is a "black box," which will not be described in detail but is expected to have the following attributes. It possesses a finite number of states, which need be *prima facie* characterized only by stating their number, say n, and by enumerating them accordingly: 1, 2, . . . , n. The essential operating characteristic of the automaton consists of describing how it is caused to change its state, that is, to go over from a state i into a state j. This change requires some interaction with the outside world, which will be standardized in the following manner. As far as the machine is concerned, let the whole outside world consist of a long paper tape. Let this tape be, say, 1 inch wide, and let it be subdivided into fields (squares) 1 inch long. On each field of this strip we may or may not put a sign, say, a dot, and it is assumed that it is possible to erase as well as to write in such a dot. A field marked with a dot will be called a "1," a field unmarked with a dot will be called a "0." (We might permit more ways of marking, but Turing showed that this is irrelevant and does not lead to any essential gain in generality.) In describing the position of the tape relative to the automaton it is assumed that one particular field of the tape is under direct inspection by the automaton, and that the automaton has the ability to move the tape forward and backward, say, by one field at a time. In specifying this, let the automaton be in the state i (= 1, . . . , n), and let it see on the tape an e (= 0, 1). It will then go over into the state j (= 0, 1, . . . , n), move the tape by p fields (p = 0, +1, −1; +1 is a move forward, −1 is a move backward), and inscribe into the new field that it sees f (= 0, 1; inscribing 0 means erasing; inscribing 1 means putting in a dot). Specifying j, p, f as functions of i, e is then the complete definition of the functioning of such an automaton.

Turing carried out a careful analysis of what mathematical processes can be effected by automata of this type. In this connection he proved various theorems concerning the classical "decision problem" of logic, but I shall not go into these matters here. He did, however, also introduce and analyze the concept of a "universal automaton," and this is part of the subject that is relevant in the present context.

An infinite sequence of digits e (= 0, 1) is one of the basic entities in mathematics. Viewed as a binary expansion, it is essentially equivalent to the concept of a real number. Turing, therefore, based his consideration on these sequences.

He investigated the question as to which automata were able to construct which sequences. That is, given a definite law for the formation of such a sequence, he inquired as to which automata can be used to form the sequence based on that law. The process of "forming" a sequence is interpreted in this manner. An automaton is able to "form" a certain sequence if it is possible to specify a finite length of tape, appropriately marked, so that, if this tape is fed to the automaton in question, the automaton will thereupon write the sequence on the remaining (infinite) free portion of the tape. This process of writing the infinite sequence is, of course, an indefinitely continuing one. What is meant is that the automaton will keep running indefinitely and, given a sufficiently long time, will have inscribed any desired (but of course finite) part of the (infinite) sequence. The finite, premarked, piece of tape constitutes the "instruction" of the automaton for this problem.

An automaton is "universal" if any sequence that can be produced by any automaton at all can also be solved by this particular automaton. It will, of course, require in general a different instruction for this purpose.

The Main Result of the Turing Theory

We might expect *a priori* that this is impossible. How can there be an automaton which is at least as effective as any conceivable automaton, including, for example, one of twice its size and complexity?

Turing, nevertheless, proved that this is possible. While his construction is rather involved, the underlying principle is nevertheless quite simple. Turing observed that a completely general description of any conceivable automaton can be (in the sense of the foregoing definition) given in a finite number of words. This description will contain certain empty passages—those referring to the functions mentioned earlier (j, p, f in terms of i, e), which specify the actual functioning of the automaton. When these empty passages are filled in, we deal with a specific automaton. As long as they are left empty, this schema represents the general definition of the general automaton. Now it becomes possible to describe an automaton which has the ability to interpret such a definition. In other words, which, when fed the functions that in the sense described above define a specific automaton, will thereupon function like the object described. The ability to do this is no more mysterious than the ability to read a dictionary and a grammar and to follow their instructions about the uses and principles of combinations of words. This automaton, which is constructed to read a description and to imitate the object described, is then the universal automaton in the sense of Turing. To make it duplicate any operation that any other automaton can perform, it suffices to furnish it with a description of the automaton in question and, in addition, with the instructions which that device would have required for the operation under consideration.

Broadening of the Program to Deal with Automata That Produce Automata

For the question which concerns me here, that of "self-reproduction" of automata, Turing's procedure is too narrow in one respect only. His automata are purely computing machines. Their output is a piece of tape with zeros and ones on it. What is needed for the construction to which I referred is an automaton whose output is other automata. There is, however, no difficulty in principle in dealing with this broader concept and in deriving from it the equivalent of Turing's result.

The Basic Definitions

As in the previous instance, it is again of primary importance to give a rigorous definition of what constitutes an automaton for the purpose of the investigation. First of all, we have to draw up a complete list of the elementary parts to be used. This list must contain not only a complete enumeration but also a complete operational definition of each elementary part. It is relatively easy to draw up such a list, that is, to write a catalogue of "machine parts" which is sufficiently inclusive to permit the construction of the wide variety of mechanisms here required, and which has the axiomatic rigor that is needed for this kind of consideration. The list need not be very long either. It can, of course, be made either arbitrarily long or arbitrarily short. It may be lengthened by including in it, as elementary parts, things which could be achieved by combinations of others. It can be made short—in fact, it can be made to consist of a single unit by endowing each elementary part with a multiplicity of attributes and functions. Any statement on the number of elementary parts required will therefore represent a common-sense compromise, in which nothing too complicated is expected from any one elementary part, and no elementary part is made to perform several, obviously separate, functions. In this sense, it can be shown that about a dozen elementary parts suffice. The problem of self-reproduction can then be stated like this: Can one build an aggregate out of such elements in such a manner that if it is put into a reservoir, in which there float all these elements in large numbers, it will then begin to construct other aggregates, each of which will at the end turn out to be another automaton exactly like the original one? This is feasible, and the principle on which it can be based is closely related to Turing's principle outlined earlier.

Outline of the Derivation of the Theorem Regarding Self-Reproduction
First of all, it is possible to give a complete description of everything that is an automaton in the sense considered here. This description is to be conceived as a general one, that is, it will again contain empty spaces. These empty spaces have to be filled in with the functions which describe the actual structure of an automaton. As before, the difference between these spaces filled and unfilled is the difference between the description of a specific automaton and the general description of a general automaton. There is no difficulty of principle in describing the following automata.

(a) Automaton A, which when furnished the description of any other automaton in terms of appropriate functions, will construct that entity. The description should in this case not be given in the form of a marked tape, as in Turing's case, because we will not normally choose a tape as a structural element. It is quite easy, however, to describe combinations of structural elements which have all the notational properties of a tape with fields that can be marked. A description in this sense will be called an instruction and denoted by a letter I.

"Constructing" is to be understood in the same sense as before. The constructing automaton is supposed to be placed in a reservoir in which all elementary components in large numbers are floating, and it will effect its construction in that milieu. One need not worry about how a fixed automaton of this sort can produce others which are larger and more complex than itself. In this case the greater size and the higher complexity of the object to be constructed will be reflected in a presumably still greater size of the instructions I that have to be furnished. These instructions, as pointed out, will have to be aggregates of elementary parts. In this sense, certainly, an entity will enter the process whose size and complexity is determined by the size and complexity of the object to be constructed.

In what follows, all automata for whose construction the facility A will be used are going to share with A this property. All of them will have a place for an instruction I, that is, a place where such an instruction can be inserted. When such an automaton is being described (as, for example, by an appropriate instruction), the specification of the location for the insertion of an instruction I in the foregoing sense is understood to form a part of the description. We may, therefore, talk of "inserting a given instruction I into a given automaton," without any further explanation.

(b) Automaton B, which can make a copy of any instruction I that is furnished to it. I is an aggregate of elementary parts in the sense outlined in (a), replacing a tape. This facility will be used when I furnishes a description of another automaton. In other words, this automaton is nothing more subtle than a "reproducer"—the machine which can read a punched tape and produce a second punched tape that is identical with the first. Note that this automaton, too, can produce objects which are larger and more complicated than itself. Note again that there is nothing surprising about it. Since it can only copy, an object of the exact size and complexity of the output will have to be furnished to it as input.

After these preliminaries, we can proceed to the decisive step.

(c) Combine the automata A and B with each other, and with a control mechanism C which does the following. Let A be furnished with an instruction I (again in the sense of [a] and [b]). Then C will first cause A to construct the automaton which is described by this instruction I. Next C will cause B to copy the instruction I referred to above, and insert the copy into the automaton referred to above, which has just been constructed by A. Finally, C will separate this construction from the system A + B + C and "turn it loose" as an independent entity.

(d) Denote the total aggregate A + B + C by D.

(e) In order to function, the aggregate D = A + B + C must be furnished with an instruction I, as described above. This instruction, as pointed out above, has to be inserted into A. Now form an instruction I_D, which describes this automaton D, and insert it into A within D. Call the aggregate which now results E.

E is clearly self-reproductive. Note that no vicious circle is involved. The decisive step occurs in E, when the instruction I_D, describing D, is constructed and attached to D. When the construction (the copying) of I_D is called for, D exists already, and it is in no wise modified by the construction of I_D. I_D is simply added to form E. Thus there is a definite chronological and logical order in which D and I_D have to be formed, and the process is legitimate and proper according to the rules of logic.

Interpretations of This Result and of Its Immediate Extensions

The description of this automaton E has some further attractive sides, into which I shall not go at this time at any length. For instance, it is quite clear that the instruction I_D is roughly effecting the functions of a gene. It is also clear that the copying mechanism B performs the fundamental act of reproduction, the duplication of the genetic material, which is clearly the fundamental operation in the multiplication of living cells. It is also easy to see how arbitrary alterations of the system E, and in particular of I_D, can exhibit certain typical traits which appear in connection with mutation, lethally as a rule, but with a possibility of continuing reproduction with a modification of traits. It is, of course, equally clear at which point the analogy ceases to be valid. The natural gene does probably not contain a complete description of the object whose construction its presence stimulates. It probably contains only general pointers, general cues. In the generality in which the foregoing consideration is moving, this simplification is not attempted. It is, nevertheless, clear that this simplification, and others similar to it, are in themselves of great and qualitative importance. We are very far from any real understanding of the natural processes if we do not attempt to penetrate such simplifying principles.

Small variations of the foregoing scheme also permit us to construct automata which can reproduce themselves and, in addition, construct others. (Such an automaton performs more specifically what is probably a—if not the—typical gene function, self-reproduction plus production—or stimulation of production—of certain specific enzymes.) Indeed, it suffices to replace the I_D by an instruction I_{D+F}, which describes the automaton D plus another given automaton F. Let D, with I_{D+F} inserted into A within it, be designated by E_F. This E_F clearly has the property already described. It will reproduce itself, and, besides, construct F.

Note that a "mutation" of E_F, which takes place within the F-part of I_{D+F} in E_F, is not lethal. If it replaces F by F', it changes E_F into $E_{F'}$, that is, the "mutant" is still self-reproductive; but its by-product is changed—F' instead of F. This is, of course, the typical non-lethal mutant.

All these are very crude steps in the direction of a systematic theory of automata. They represent, in addition, only one particular direction. This is, as I indicated before, the direction towards forming a rigorous concept of what constitutes "complication." They illustrate that "complication" on its lower levels is probably degenerative, that is, that every automaton that can produce other automata will only be able to produce less complicated ones. There is, however, a certain minimum level where this degenerative characteristic ceases to be universal.

At this point automata which can reproduce themselves, or even construct higher entities, become possible. This fact, that complication, as well as organization, below a certain minimum level is degenerative, and beyond that level can become self-supporting and even increasing, will clearly play an important role in any future theory of the subject.

The Heart of the Proof

Chapter 5 sketches our proof that Darwinian evolution works in our toy model, but omits some crucial details, since most people will be satisfied with an approximate idea of the proof. For those who want to know more, we now outline these missing crucial details. This is once more a proof sketch, but this time it is sufficiently detailed that it becomes a routine matter for anyone with an expert knowledge of algorithmic information theory to complete the proof.

This is called "presenting a proof by successive approximations," and in my opinion it is better than throwing all the details at someone right away. This is also closer to the way that a proof is actually found when one is doing mathematical research.

The halting probability Ω is defined as follows. Each K-bit program that halts contributes $1/2^K$ to the halting probability:

$$\Omega = \sum_{\text{program P halts}} 2^{-(\text{the size in bits of P})}$$

For this sum to be < 1 instead of diverging to infinity, the programs P must be "self-delimiting," so that no extension of a valid program P is a valid program.

Next we define Ω_K to be that part of the sum defining Ω that includes only those programs up to K bits in size that halt within time K:

$$\Omega_K = \sum_{\text{program P is} \leq \text{K bits in size and halts in time} \leq \text{K}} 2^{-(\text{the size in bits of P})}$$

Ω_K is a computable function of K which converges (very slowly) to Ω in the limit from below. Furthermore,

$$\Omega_K \leq \Omega_{K+1},$$

Ω_K is a "monotone increasing" (*i.e.*, non-decreasing) function of K.

Note that knowing a very large integer is equivalent to knowing a very good lower bound on Ω:

1. If we are given an extremely large N we can find a good lower bound on Ω, namely Ω_N.
2. Conversely, given a number $\beta < \Omega$ that is greater than zero and which has a finite base-two representation (these are called dyadic fractions), we can convert β into an extremely large integer N by finding the first N for which $\Omega_N \geq \beta$. This process will never halt if $\beta > \Omega$. (It is impossible for β to equal Ω because Ω is provably irrational.)

In this manner one can see—we omit the details—that knowing the extremely large integer BB(N) is more or less equivalent to knowing N bits of Ω.

We are finished with the preliminaries.

The heart of the proof resides in considering mutations M_K which are in effect given a lower bound on Ω and which attempt to improve it (bring it closer to Ω without surpassing the value of Ω) by adding $1/2^K$ to it.

Let us denote the fitness of an arbitrary organism/program A by φ_A: this is the positive integer that A calculates before halting. Now for a complete explanation of how M_K works.

The mutation M_K, if it is given the organism A with fitness φ_A, will first of all run the program A to calculate its fitness φ_A and then will use this very big integer to calculate a lower bound Ω_{φ_A} on Ω. Then M_K adds $1/2^K$ to this lower bound on Ω obtaining

$$\beta = \Omega_{\varphi_A} + 1/2^K$$

which may be less than Ω or may surpass Ω in value. In either case, the mutation M_K converts A into the mutated program

$$A' = (\text{the prefix } \pi \text{ concatenated with } \beta) = \pi\,\beta.$$

The self-delimiting prefix π converts β into a large positive integer by finding the first N for which

$$\Omega_N \geq \beta$$

then π outputs that integer N and halts.

Why are these mutations M_K so useful?

First of all, note that if β is less than Ω, then the fitness of A' will be greater than that of A. If on the other hand β has surpassed the value of Ω, then the mutated organism A' will fail to halt.

These mutations M_K can be used to quickly converge on the correct value of Ω. If we start with A = π β with β having the value zero and we try mutation M_K for K = 1, 2, 3 . . . we will obtain a new bit of Ω at each stage, one bit per mutation tried. For if M_K succeeds, then the Kth bit of Ω is a 1; whereas if M_K fails, then the Kth bit of Ω is a 0.

Therefore the β in the program π β will have N bits of Ω correct after trying each M_K with K = 1, 2, 3 . . . N. Furthermore after these N mutations have been tried the fitness of the program π β will be approximately BB(N), which grows extremely quickly. In fact it is not difficult to see that this is essentially the best one can do after trying N mutations.

This is what we are calling "intelligent design," because we get to pick the mutations to be tried, and we can do this in the best possible order . . .

But what happens if mutations are picked at random? Then every possible mutation will be tried infinitely often, including the mutations M_K with every possible K. The presence of other mutations does no harm; as soon as we have tried M_K for K = 1, 2, 3 . . . N in order, the fitness will reach BB(N) as before. And as we showed in Chapter 5, this will happen when we have tried somewhere between N^2 and N^3 random mutations, *QED.*

Further Reading

Maurício Abdalla, *La crisis latente del darwinismo,* Cauac Editorial Nativa, 2010.

David Berlinski, *Black Mischief: The Mechanics of Modern Science,* William Morrow, 1986.

David Berlinski, *The Devil's Delusion: Atheism and Its Scientific Pretensions,* Crown Forum, 2008.

Sydney Brenner, *My Life in Science,* Biomed Central Ltd., 2001.

William Byers, *The Blind Spot: Science and the Crisis of Uncertainty,* Princeton University Press, 2011.

Gregory Chaitin, "Randomness in Arithmetic and the Decline and Fall of Reductionism in Pure Mathematics," in John Cornwell, *Nature's Imagination: The Frontiers of Scientific Vision,* Oxford University Press, 1995, pp. 27–44.

Gregory Chaitin, *Meta Math! The Quest for Omega,* Pantheon, 2005.

Gregory Chaitin, *Thinking about Gödel and Turing: Essays on Complexity, 1970–2007,* World Scientific, 2007.

Gregory Chaitin, *Matemáticas, Complejidad y Filosofía / Mathematics, Complexity and Philosophy,* Midas, 2011 (bilingual Spanish/English edition).

Gregory Chaitin, Newton da Costa and Francisco Antonio Doria, *Gödel's Way: Exploits into an Undecidable World,* CRC Press, 2012.

Martin Davis, *The Undecidable: Basic Papers on Undecidable Propositions, Unsolvable Problems and Computable Functions,* Dover, 2004.

Richard Dawkins, *The Ancestor's Tale: A Pilgrimage to the Dawn of Evolution,* Houghton Mifflin, 2004.

James Gleick, *The Information: a History, a Theory, a Flood,* Pantheon, 2011.

Ernst Haeckel, *Art Forms from the Ocean,* Prestel, 2009.

Ernst Haeckel, *Art Forms in Nature,* Prestel, 2010.

Fred Hoyle, *The Intelligent Universe: A New View of Creation and Evolution,* Michael Joseph, 1983.

John Kemeny, "Man Viewed as a Machine," *Scientific American* **192** (April 1955), pp. 58–67.

Julien Offray de La Mettrie, *Man a Machine,* Open Court, 1912.

Stephen C. Meyer, *Signature in the Cell: DNA and the Evidence for Intelligent Design,* HarperOne, 2009.

Edward F. Moore, "Artificial Living Plants," *Scientific American* **195** (October 1956), pp. 118–126.

Matt Ridley, *The Red Queen: Sex and the Evolution of Human Nature,* Harper Perennial, 2003.

Matt Ridley, *Francis Crick: Discoverer of the Genetic Code,* Eminent Lives, 2006.

Máximo Sandín, *Lamarck y los mensajeros: la función de los virus en la evolución,* Ediciones Istmo, 1995.

Máximo Sandín, *Pensando la evolución, pensando la vida: la biología más allá del darwinismo,* Cauac Editorial Nativa, 2010.

Arturo Sangalli, *Pythagoras' Revenge: A Mathematical Mystery,* Princeton University Press, 2009.

James A. Shapiro, *Evolution: A View from the 21st Century,* FT Press, 2011.

Neil Shubin, *Your Inner Fish: A Journey into the 3.5-Billion-Year History of the Human Body,* Pantheon, 2008.

John Maynard Smith, *The Problems of Biology,* Oxford University Press, 1986.

John Maynard Smith, *Shaping Life: Genes, Embryos and Evolution,* Weidenfeld & Nicolson, 1998.

John Maynard Smith and Eörs Szathmáry, *The Origins of Life: From the Birth of Life to the Origin of Language,* Oxford University Press, 1999.

John C. Stillwell, *Roads to Infinity: The Mathematics of Truth and Proof,* A K Peters, 2010.

John von Neumann, "The General and Logical Theory of Automata," in Lloyd Jeffress, *Cerebral Mechanisms in Behavior: The Hixon Symposium,* John Wiley and Sons, 1951, pp. 1–41. Reprinted in James R. Newman, *The World of Mathematics,* Dover, 2003.

John von Neumann, *Theory of Self-Reproducing Automata,* University of Illinois Press, 1966 (edited and completed by A. W. Burks).

Index

Printed in the United States
by Baker & Taylor Publisher Services